Michael Frauchiger (Ed.)

Modalities, Identity, Belief, and Moral Dilemmas

Lauener Library of Analytical Philosophy

Edited by
Wilhelm K. Essler, Dagfinn Føllesdal,
and Michael Frauchiger

Volume 3

Modalities, Identity, Belief, and Moral Dilemmas

Themes from Barcan Marcus

Edited by
Michael Frauchiger

DE GRUYTER

ISBN 978-3-11-043858-1
e-ISBN (PDF) 978-3-11-042955-8
e-ISBN (EPUB) 978-3-11-042972-5
ISSN 2198-2155

Library of Congress Cataloging-in-Publication Data
A CIP catalog record for this book has been applied for at the Library of Congress.

Bibliographic information published by the Deutsche Nationalbibliothek
The Deutsche Nationalbibliothek lists this publication in the Deutsche Nationalbibliografie; detailed bibliographic data are available on the Internet at http://dnb.dnb.de.

Printing and binding: CPI books GmbH, Leck
∞ Printed on acid-free paper
Printed in Germany

MIX
Papier aus verantwortungsvollen Quellen
FSC® C083411
FSC www.fsc.org

www.degruyter.com

Modalities, Identity, Belief, and Moral Dilemmas

Themes from Barcan Marcus

The contributions to this collection, which hark back to the 3rd International Lauener Symposium in honour of Ruth Barcan Marcus, open up originary and stimulating perspectives by leading thinkers on a broad variety of Barcan Marcus's concerns, ranging from the systematic foundation and interpretation of quantified modal logic, identity and indiscernibility, nature of extensionality, intensional languages, necessity of identity, direct reference theory for proper names, notions of essentialism, second-order modal logic, modal metaphysics, properties, classes and assortments, substitutional and objectual interpretations of quantification, actualism, the Barcan formula, possibilia and possible-world semantics to epistemic and deontic modalities, states of affairs, non-language-centered theories of belief, theories of rationality, consistency of a moral code, moral dilemmas, and much more. The fully revised and reworked contributions to this volume are critically directed toward various aspects of thorough Marcusian approaches and argumentations, and they demonstrate that Barcan Marcus's highly original and clear ideas have had a formative, determining influence on the direction in which certain themes central to today's philosophical debate have developed. Further, the collection includes an orientating, insightful and detailed intellectual autobiography from Ruth Barcan Marcus herself as well as an informal interview with her containing her unfiltered, frank and open answers, both of them revealing impressive facets of the rich life and the keen, honest thinking of this extraordinary philosopher and courageous person. The book brings together contributions by Ruth Barcan Marcus, Timothy Williamson, Dagfinn Føllesdal, Joëlle Proust, Pascal Engel, Edgar Morscher, Erik J. Olsson, Michael Frauchiger.

Contents

Michael Frauchiger

Proem: Highlighting Ruth Barcan Marcus's Courageous Philosophical Life and Work

The present collection on themes from Barcan Marcus had its source in the 3rd International Lauener Symposium held in Bern, Switzerland, in 2008, during which the Lauener Prize for an Outstanding Oeuvre in Analytical Philosophy for that year was presented to Ruth Barcan Marcus at a special award ceremony. The event was sponsored by the Lauener Foundation, which had been founded and endowed by the Swiss philosopher Henri Lauener (1933–2002), a longstanding academic friend of Ruth Barcan Marcus's and the organizer of a memorable, much-valued series of international Philosophy Colloquia in Biel and Bern, in which Barcan Marcus had been involved at one time. This special constellation enabled the Lauener Foundation to realize in Bern a demanding symposium on themes from Barcan Marcus in which she herself participated actively, even though at this point she had not traveled abroad any more for a decade[1].

Henri Lauener had much respect for Ruth Barcan Marcus as a pioneer of quantified modal logic and its philosophical reflection and, moreover, for her line of persistent, refined resistance to W. V. Quine's acute retorts to her systematic contributions to improving the formal structure, semantics, ontology, and thus intelligibility, of modal logic. Lauener – despite being critical of important aspects of Quinean philosophy (in particular of strongly naturalist ones) – considered Quine the leading contemporary American philosopher and one of the subtlest and, at the same time, deepest thinkers in the history of modern philosophy.[2] In comparison, Barcan Marcus's reactions to the longstanding controversy in which she became entangled with Quine concerning modal logics reflect a somewhat more ambivalent appreciation of Quine the philosopher and the man. For one thing, Barcan Marcus acknowledged that Quine's early [I quote] "criticisms and the continuing debate were a catalyst for some of my subsequent work."[3] For another thing, she points out annoying parts of Quine's continuous countermoves to her own logical and philosophical

Lauener-Stiftung / Lauener Foundation for Analytical Philosophy

1 Cp. Barcan Marcus (2010), 90. (Also, this volume, 37.)
2 Cp. Lauener (1982), 11.
3 Barcan Marcus (1993), x.

work. In the present volume, both Barcan Marcus's autobiographical piece "A Philosopher's Calling" and the interview bear witness of her displeasure at Quine's animadversion against her pioneering systematic studies of quantified modal logic, as expressed e.g. in the following passage from "A Philosopher's Calling", which indicates some concern about an inclusive attitude towards opponents in cooperative inquiry: "Quine's negative views had been expressed immediately on the heels of my publications (...) It was as if such efforts needed to be nipped in the bud. (...) My name is Ruth and I was in alien corn".[4]

It's worth noting, though, that there were not only points of divergence, but some points of agreement as well in the course of this long-running debate between Barcan Marcus and Quine. For instance, both of them reject (mere) possibilia, i.e. merely possible, nonactual objects, though the details of their respective justifications of the critique of possibilia differ considerably. – In accordance with Barcan Marcus's causal, or historical-chain, account of direct reference for genuine proper names (limiting ourselves here to Barcan Marcus's reasons for her rejection of possibilia), such putative merely possible objects are not (empirically) encounterable and thus cannot be objects of reference and cannot be genuinely named at all. In line with this, Barcan Marcus defends an actualistic modal semantics with objectual quantification[5] and the domains (of

4 Barcan Marcus (2010), 85. (Also, this volume, 29f.) – In this passage, Barcan Marcus tells of an almost legendary colloquium at Harvard in February 1962, of which she was apprehensive at the time. In Quine's sphere, she felt like her biblical namesake, standing in alien corn. Barcan Marcus there presented her paper "Modalities and Intensional Languages", in conjunction with comments by Quine. Participating in the discussion following Barcan Marcus's lecture were (in addition to Quine and Barcan Marcus) Kripke, Føllesdal and McCarthy. (Cp. Barcan Marcus (1993), 3, 222f.) – It is important to bring in, at this point, the perspective of Dagfinn Føllesdal, who had in the previous year submitted his Ph.D. thesis to the Harvard Department of Philosophy. Føllesdal's thesis advisor was Quine and yet Barcan Marcus's innovative work in quantified modal logic was carefully and accurately discussed in his dissertation. Føllesdal's recollections of the Harvard colloquium lecture by Barcan Marcus do not justify Ruth Barcan Marcus's apprehensiveness of a lack of respect and inclusiveness for philosophical opponents within Quine's sphere of influence at Harvard in those days. Towards the end of his contribution to this volume, "Ruth Marcus, Modal Logic and Rigid Reference", Føllesdal writes: "Ruth describes in her book *Modalities* the discussion at Harvard in 1962 as if she were in a lion's den, where she appreciated Saul Kripke's support. She clearly believed that being a student of Quine I sided with Quine in his rejection of the modalities. (...) She might have been relieved in 1962, when she visited Harvard, if she had known that she had one more ally in that group". (This volume, 47f).

5 At many points in her anthology of collected essays entitled *Modalities*, Barcan Marcus advocates a substitutional semantics for quantification. At one point, however, she acknowledges that objectual interpretation of quantification remains ultimately indispensable for

individuals) of the alternative possible worlds coextensive with or included in the domain of this actual world, hence validating the Barcan formula and ruling out possibilia.[6] – Getting to the heart of her critique and rejection of possibilia and, in addition, encapsulating "the central truth in Quine's critique of possibilia"[7], Barcan Marcus writes: "It seems harmless in formal semantics to speak of assigning an object from this or any other world to a variable or to a name. But we are in this actual world, users of our actual language. (...) Naming relates a word introduced into an actual language in the actual world to a thing that is there to be encountered in the world when the event of naming occurs. (...) That one has no general criteria of identity for possibilia is not sufficient for rejecting them. As we noted, even general criteria of identification for actual material objects seem, also, to elude us. (...) It is not the absence of criteria that makes us dubious. It is rather that what is absent is the individuals. They are not there to be objects of reference at all. (...) There are no individual objects, which are what is required for an identity relation. There are no traceable histories, origins, futures, and so on. Criteria aside, Quine is correct when he says, "The concept of identity is inapplicable to 'unactualized possibles'." No identity, no entity."[8]

Notwithstanding the varying pitch of their reactions to Quine's giant philosophical shadow, Barcan Marcus and Lauener looked at each other with much mutual appreciation and inquisitiveness. Henri Lauener respected Ruth Barcan Marcus not only for her acute and subtle mind, but also for her personal courage, and he associated Barcan Marcus's remarkable courage with her dogged commitment to integrity and honesty[9]. I feel it is appropriate at this point to add that in the academic milieu – and probably in the world of work or

possible-world semantics; for, even if substitutional quantifiers are used in quantified modal logic, objectual quantification is also required in the final analysis. On p. 213 of Barcan Marcus (1993), Marcus writes with regard to substitutional possible-world semantics: "Identity, which is a feature of *objects*, cannot be defined in such a semantics. Intersubstitutivity of *syntactical* items *salva veritate* does not generate objects, which must be *given* if identity is to hold. (...) Substitutional semantics may have some uses for nonobjectual discourse, but, as I now believe, only in conjunction with objectual quantification for the domain of actuals." – So, on a substitutional interpretation of quantifiers, which does not assign any domains to possible worlds and, accordingly, no objects to names, the relation between linguistic expressions and actual objects in the world is neglected and left out of account.

6 Cp. e.g. Barcan Marcus's illuminating paper "Possibilia and Possible Worlds", in Barcan Marcus (1993), 189–213.

7 Barcan Marcus (1993), 204.

8 Barcan Marcus (1993), 207f.

9 Cp. Lauener (1999), 177.

rather in the day-to-day realities of life generally – it takes indeed a lot of courage and resistance, on the part of women in particular, to maintain high levels of autonomy and genuineness. This is reflected in many pertinent passages throughout Barcan Marcus's autobiographical piece reprinted in this volume, a few of which are cited here: "We were now a family of four females, which I mention because I believe I had an easier time following an unconventional path than if there had been a strong male presence in the family."[10] – "There were recurrent discriminatory episodes. I'll mention one graphic incident among many. Yale had a philosophy club open to undergraduate and graduate students. I was elected president but then received a letter from the chair of the department suggesting that I decline. The reasons given were that Yale was predominantly and historically a male institution and that my election may have been a courtesy. Also, the club's executive committee met at Mory's which was closed to women. I did not respond to the letter and did not decline. It was, to me, obviously unreasonable. When the letter was discussed in the graduate dining room, several students said it was imprudent to have revealed it since the chair influences job placement. My response was as before: I wasn't thinking about a job. I assumed the presidency and the executive committee did *not* meet at Mory's (which was not "liberated" until the seventies)."[11] – "It seems I was not a unanimous choice of the search committee. The protest was that there never had been a woman chair at the University of Illinois! But it happened, and I embarked on one of my careers as a proper "professional."[12] – "But I am essentially a loner. One of the changes in academic style in recent years is the distribution of papers by an author for comment by large, sometimes astonishingly large, numbers of contemporaries, which is then noted in the acknowledgments. That was not my style. There was often no point, in any case, since I characteristically defended positions contrary to received views, if there were received views."[13]

At last, Lauener's high respect for Ruth Barcan Marcus's inventive, accurate, far-sighted philosophical oeuvre resulted in a *Festschrift* for her[14], which he edited and published in *Dialectica* (a journal of philosophy that was originally founded in 1947 by Ferdinand Gonseth, Paul Bernays and Gaston Bachelard and which Henri Lauener had been editing for over twenty years at that point in

10 Barcan Marcus (2010), 76. (Also, this volume, 19.)
11 Barcan Marcus (2010), 80f. (Also, this volume, 24.)
12 Barcan Marcus (2010), 86. (Also, this volume, 31f.)
13 Barcan Marcus (2010), 90. (Also, this volume, 36.)
14 Lauener (1999).

time, establishing it as an international platform for contributions to analytical, accurate philosophy). – Lauener's appreciation of Ruth Barcan Marcus's lucid and forceful writing in sparing doses that have a strong, positive effect is shared by many, and her most elaborate papers thus often serve as models for incisive philosophical writing. Barcan Marcus herself comments on her writing thus: "I am not driven to publish. Papers are submitted where I think I have a useful account of or solution to a clear question of logical or philosophical interest. The questions usually originate in some common sense observations, couched in our common, ordinary language."[15] Always an unbending critic of her own work, Ruth Barcan Marcus wrote with exceptional prudence and her continuing [I quote] "worry about the best way to make my exposition clearer"[16] sometimes led to a considerable deferral of the publication[17] – or even to the complete withdrawal[18] – of certain papers she had already completed previously. Barcan

15 Barcan Marcus (2010), 82. (Also, this volume, 26.)

16 Cp. Barcan Marcus (1993), 89.

17 For example, Barcan Marcus's remarkable paper "Classes, Collections, Assortments, and Individuals" (Barcan Marcus (1993), 89–100) had quite a protracted publication history. Marcus submitted it in 1965, but it was not published until 1974, when the journal's [I quote] "patient editor presented me with an ultimatum" (p. 89). The delay was due inter alia to a prolonged [I quote] "struggle with arriving at a salient vocabulary for the distinctions I wanted to make among notions that are often conflated in varying ways, such as attribute, class, collection, set, and what I have in the present paper called "assortment."" (p. 89). This struggle for improved terminology indeed goes back to Marcus's contribution to the influential 1962 Helsinki *Colloquium on Modal and Many-Valued Logics*. And that contribution of Barcan Marcus's was in turn based on her early, pivotal studies of quantified modal logic; that is, her colloquium contribution was [I quote] "an application of "A Functional Calculus of First Order Based on Strict Implication," Journal of Symbolic Logic, XI (1946) and "The Identity of Individuals in a Strict Functional Calculus," XII (1947)" (p. 97). – In the relevant Helsinki Colloquium paper, published in 1963, as well as in the abbreviated and improved 1974 and 1993 versions of it, Barcan Marcus has developed a still thought-provoking modal theory of classes, which takes into account that assortments, unlike classes, are not given by defining conditions for membership, but by inventory. She points out that a bracketed list of tags which designates an assortment does – analogously to the tags (i.e. genuine proper names of physical objects) themselves – have a purely referential (i.e. not a predicative) function. Equivalent assortments are thus strictly identical, whereas classes may be contingently equivalent and may differ in modal contexts.

18 Barcan Marcus was invited to contribute to each of the two *Festschriften* for Henri Lauener, which were published in 1993 in *Grazer Philosophische Studien* and in 1995 in *Dialectica* respectively. Yet, whereas Barcan Marcus's well-known article "The Anti-Naturalism of Some Language Centered Accounts of Belief" appeared in the latter (Furrer/Hottinger (1995), pp. 113–129), she withdrew her already completed contribution to the former at the last minute because she considered it "not sufficiently worthy" (cp. "Vorbemerkung/Preface" of Burri,

Marcus, therefore, did certainly not believe in streamlined publication planning. Clearly, she took a distinctive, individual path to her original publications which subsequently have advanced the philosophical debate. – And on a more general note, Barcan Marcus did not believe in personal life planning. In her lively extemporized autobiographical talk at the Lauener Symposium in her honour, Ruth Marcus said: "I remember being in a legal theory workshop (...) going on about life's plan. (...) I bet a lot of people here have a life plan. I never had a life plan. I just sort of did what I wanted to do from day to day. Maybe I had a plan for the next week, or for the next month; but the odd view of having a life plan ... So I said (...): "I don't have this huge projection of my future." – "You don't?!" – I was amazed. And there was a tremendous discrepancy in the audience between those people who sort of have a life and then those who had a – plan."[19]

Barcan Marcus is probably best known as a logician and a philosopher of logic in the analytic tradition who has critically and productively improved the existing approaches in the development of modal logic and its philosophical interpretation throughout her life. Barcan Marcus was not, however, confined to the philosophy of logics related to the logician's interest she had in modal, deontic and other intensional formal languages[20]. Her philosophical thought soon spilled over to include metaphysics, epistemology, and moral philosophy. – Thus she was, for instance, very influential in the development of a revisionary conception of beliefs as cognitive attitudes to possible states of affairs (rather than relations to linguistic or quasi-linguistic entities). Such a non-language-centered account of belief apparently offers various advantages over the still dominant language-centered accounts of belief. For instance, it *disallows* to grant beliefs exclusively to language users (and to deny beliefs to

A./Freudiger, J., eds.,(1993) *Relativism and Contextualism: Essays in Honor of Henri Lauener*, in: *Grazer Philosophische Studien* 44 (1993).).

19 This is my transcript of what Ruth Barcan Marcus said about *life plan* during her talk titled "Recollections" at the 3rd International Lauener Symposium on Analytical Philosophy in honour of herself (on 30 May 2008 in Bern). At this point in her talk, Barcan Marcus told of a legal theory workshop in which she had participated along with Ronald Dworkin and others, plus an audience.

20 In her path-breaking paper "Modalities and Intensional Languages" (Barcan Marcus (1993), 3–23), Marcus discusses, inter alia, the issue of identity in intensional contexts, beginning with a definition of the notion of an intensional language, which remains illuminating up to the present day by clearly setting out degrees of intensionality: "A language is explicitly intensional to the degree to which it does not equate the identity relation with some weaker form of equivalence." (p. 5).

animals and babies thereby) or to ascribe to agents beliefs in the impossible. At the same time it *allows* to draw a parallel between belief and knowledge[21] and to divorce beliefs from acts of sincere verbal assent (thus accommodating, e.g., the possibility of unconscious beliefs) and, moreover, to link the account of the rationality of belief with a wider, more general account of the rationality of action.[22] – Furthermore, Barcan Marcus's contributions to moral philosophy and deontic logic have gained considerable attention and impact. Most notably, Barcan Marcus maintains, *pace* Kant, that moral obligations can and do conflict in reality. She holds that moral dilemmas arise when obligations can in practice not be jointly fulfilled owing to contingent circumstances. And she argues that such dilemmas are real (not merely apparent) and usually do not indicate inconsistency of the relevant set of moral principles nor inconsistency of the particular moral judgments that originate from those principles. This claim is underpinned by the definition of consistency for a moral code proposed by Barcan Marcus: a set of moral principles is consistent if and only if there is some possible world in which those principles are all obeyable in all circumstances in that world. On this basis, Barcan Marcus makes the point that the recognition of the reality of ethical dilemmas (and also of the appropriateness of attendant feelings of regret, remorse or guilt) ought to motivate and reinforce the striving of rational agents to avoid, or minimize, such moral conflicts by trying their best to bring about circumstances in which all their obligations can in future be jointly fulfilled.[23]

Through her originative work in these diverse philosophical areas – work which is characterized by her unremitting concern about the clarity of her concepts, about the reliability of her evidence (which requires due consideration of

---

21 Barcan Marcus's following statement puts this in a nutshell: "Briefly, a proper object of believing is a possible state of affairs, and a proper object of knowing is an actual state of affairs." (Barcan Marcus (1993), 153).

22 Cp. e.g. Barcan Marcus's innovative paper "Some Revisionary Proposals about Belief and Believing" (Barcan Marcus (1993), 233–255). – A more general account of the rationality of action presumably amounts to a wide-ranging view of the rationality of an agent in his diverse actions, i.e. in his activity on the whole, which requires quite a strong notion of coherence (far beyond logical consistency) of all the agent's verbal and nonverbal acts. The wider notion of rationality of action which Barcan Marcus calls for indeed goes beyond the deductive or inductive relations among sentences or (quasi-linguistic) propositions assented to, that is, beyond logical relations, in terms of which language-centered theories of belief tend to define their unduly narrow notion of rationality, which, according to Barcan Marcus, lacks explanatory force.

23 Cp. esp. Barcan Marcus's influential paper "Moral Dilemmas and Consistency" (Barcan Marcus (1993), 125–141).

results of research in other disciplines) as well as about the soundness of her argumentation – Ruth Barcan Marcus has exemplified and advanced the value of analytical, accurate, clear philosophizing.

Acknowledgements

I wish to thank the authors (Ruth Barcan Marcus, Pascal Engel, Dagfinn Føllesdal, Edgar Morscher, Erik J. Olsson, Joëlle Proust, and Timothy Williamson), who have generously contributed their thoroughly reworked papers to this anthology, for their persevering commitment to the long-winded project of this book. Also, many thanks to Wilhelm K. Essler and Dagfinn Føllesdal, my fellow coeditors of the book series "Lauener Library of Analytical Philosophy" (edited on behalf of the Lauener-Stiftung) as well as to my fellow members of the Foundation Council of the Lauener-Stiftung (Dagfinn Føllesdal, Stephan Hottinger, Alex Burri, Dale Jacquette, and Dieter Jordi) for their encouragement. I owe gratitude as well to Diana Raffman, who accompanied Ruth Barcan Marcus on her arduous journey to Bern, thereby facilitating Barcan Marcus's active participation in the Lauener Symposium in honour of her own philosophical oeuvre. Further, sincere thanks are due to Dawn Jacob, who in the last stage of Barcan Marcus's life was her ad-hoc assistant, and who was helping me with remaining in contact with Ruth Barcan Marcus until shortly before her demise. My thanks also to Rafael Hüntelmann and to Gertrud Grünkorn, Maik Bierwirth and Florian Ruppenstein, at De Gruyter, for their support and cooperation in guiding the book to final publication at long last.

Finally, I am deeply grateful to Ruth Barcan Marcus herself for the spirit of collaboration she showed with regard to the preparations for the Symposium in her honour and with respect to the plans for the present anthology on themes from her as well as, beyond this, for her optimism and goodwill even when she was in poor health towards the end of her days. – At the end of her autobiographical piece reprinted in this book, Barcan Marcus dedicates a passage to her participation in the Lauener Foundation's 2008 Symposium honouring her oeuvre, where she notes: "There were challenging papers, to which I have still to respond."[24] In fact, all the papers originally presented during the Symposium have thereafter been thoroughly revised and, in part, completely reworked and expanded for inclusion in the present book. All in all, this entire reworking process extended over a period of several years. Initially Ruth Barcan Marcus's

24 Barcan Marcus (2010), 90. (Also, this volume, 37.)

intention was to respond to all the final versions of papers in one detailed, long commentary. But after a time she wrote me that her optimism of recovering sufficiently to write the in-depth responses which the very interesting papers deserved was unwarranted. Ultimately Barcan Marcus was still eager to frame brief, sketchy responses to those finalized papers she had received by then, and she actually gave her responses to those contributions thought, but her illness persisted and her frail physical health remained a severe impediment, so that she did not have opportunity to write down her thoughts in even a sketchy manner (as has been confirmed to me by Barcan Marcus's assistant). About a quarter of a year before she died, in the last e-mail which I received from her personally, Ruth Barcan Marcus wrote: "I have not been well, but I am delighted that the book will be coming out. Thank you for your patience".

References

Barcan Marcus, R.(1993) *Modalities: Philosophical Essays*, New York, Oxford: Oxford University Press.

Barcan Marcus, R.(2010) "A Philosopher's Calling", in: *Proceedings and Addresses of the American Philosophical Association* 84, no. 2 (November 2010), 75–92. Reprinted in this volume, 17–37.

Furrer, S./Hottinger, S., eds., (1995) *Transcendentalism or Naturalism: Proceedings of the International Congress in Honour of Prof. Henri Lauener's 60th Birthday in Berne, October 14–16, 1993*, in: *Dialectica* 49, no. 2–4 (1995).

Lauener, H. (1982) *Willard Van Orman Quine*, München: C.H.Beck.

Lauener, H., ed., (1999) *Festschrift zu Ehren von Ruth Barcan Marcus*, in: *Dialectica* 53, no. 3–4 (1999).

Timothy Williamson

Laudatio: Ruth Barcan Marcus (1921–2012)

The central methodological advantage that analytic philosophy enjoys over all other forms of philosophy, past and present, is the rigorous framework of formal logic within which it can conduct its inquiries. Although different systems of logic are needed for different branches of philosophical inquiry, in the core area of metaphysics and surrounding fields for the past forty years the most natural and fruitful setting for inquiry has been *quantified modal logic*, in which we not only have the resources of first-order logic with identity but can also raise explicit questions of possibility and necessity with elegantly perspicuous generality.

The first published study of quantified modal logic as a branch of formal logic appeared in March 1946 in *The Journal of Symbolic Logic*, under the title 'A functional calculus of first order based on strict implication', by Ruth C. Barcan, a logician whose identity with Professor Marcus is of course necessary. The system that she presented there did not simply combine pre-existing non-modal quantified logics with pre-existing unquantified modal logics. It identified a crucial axiom about the interaction of the two sides, the interchange of modal operators with quantifiers. The axiom says that if there *can* be something that has a certain property, then there *is* something that can have that property. This is the famous Barcan formula; most logicians can only dream of having a formula named after them. Its converse is also derived in the paper. The Barcan formula and its converse are neither obviously correct nor obviously incorrect (on the intended interpretation), but they are of the utmost importance, both technical and philosophical, to the distinctive nature of quantified modal logic. Technically, their presence or absence makes a large strategic difference to the ways in which the proof theory and formal semantics of quantified modal systems can be developed. But this is closely connected to their philosophical significance too, for together they are tantamount to the claim that it is non-contingent what individuals there are. Although that non-contingency claim may sound implausible on first hearing, it can be given a sustained defence in more than one way, either by taking a narrow view of what individuals there can be or by taking a broad view of what individuals there are. In metaphysics there are disputes whose content is notoriously hard to pin down, for example

University of Oxford

concerning actualism (the thesis that 'everything is actual') and its analogue for time, presentism (the thesis that 'everything is present'). These disputes are threatened by trivialization; they can easily sound verbal. It is increasingly appreciated that the best way to focus them on worthwhile issues may be to reconfigure them as disputes over the validity of the Barcan formula and its converse and their analogues in tense logic. Those formulas lie at the heart of other metaphysical debates too: for example, they present a lethal threat to one contemporary version of the correspondence theory of truth, according to which a truth has to be made true by some *thing*, a truthmaker. We are going to be hearing much, much more of the Barcan formula and its converse in metaphysics.

The 1946 paper did not initially meet with universal acclaim, although its importance was recognized by C.I. Lewis, one of the founders of modern modal logic. Its main critic was W.V.O. Quine, who argued that its application of modal operators to formulas with free variables was incoherent and unintelligible (although he did concede that Miss Barcan 'is scrupulous over the distinction between use and mention of expressions – a virtue rare in the modality literature': rare praise from the guardian of the use-mention distinction). Quine's original criticisms were technically unsound, and he was forced over the years into a series of revisions that eventually reduced the charge to one of a commitment to Aristotelian essentialism. Even there, technical results vindicated Professor Marcus's later reply that the commitment was to the *intelligibility*, not the *truth*, of essentialism, and that in any case there may well be a scientific basis for some form of essentialism. Philosophy has gone Marcus's way, not Quine's, but the vindication of her paper was a gradual process: it was years ahead of its time.

In 1947, Miss Barcan published another pioneering paper on quantified modal logic, 'The identity of individuals in a strict functional calculus of second order'. It is best known for the first proof of the necessity of identity, the thesis that individuals cannot be contingently identical. For many years this was regarded as a paradox, perhaps even a *reductio ad absurdum* of quantified modal logic. There were thought to be obvious examples of contingent identity. But on further analysis the apparent counterexamples turned out to rely on philosophical confusions, concerning either the scope of definite descriptions or the distinction between the contingent and the *a posteriori*, or at least on deeply questionable metaphysical assumptions. In contemporary philosophy, the necessity of identity is widely seen as a vital insight into modal metaphysics, and a valuable constraint on philosophical theorizing.

The 1947 paper is pioneering in another respect too. It is a system of *second-order* modal logic. That is, it permits quantification into predicate position, not just into name position as in first-order logic. In cruder terms, it lets one generalize over properties, not just over the individuals that have those properties. Despite Quine's opposition, second-order non-modal logic is now widely recognized as the appropriate logical framework for many mathematical theories and other applications. But very little attention has been paid to second-order modal logic. I predict that it will play an increasingly central role as the framework for many debates in metaphysics and other areas of philosophy, and that this aspect of the 1947 paper will turn out to have been more than sixty years ahead of its time.

Who was the author of these seminal works? She was born on 2nd August 1921 in New York, to Sam and Rose Barcan, Russian Jewish immigrants who had settled in the Bronx. Ruth Barcan graduated in 1941 with a B.A. in mathematics and philosophy from New York University, where she had also studied some physics, history and classics. Informally, she learnt mathematical logic there from J.C.C. McKinsey, who encouraged her precocious interest in both technical and philosophical aspects of modal logic and her move to Yale for graduate studies. There she received her Ph.D. in 1946 with a dissertation on quantified modal logic, supervised by Frederic Fitch. Her early papers were the fruits of that research. She spent the year 1947–8 as a postdoctoral fellow at the University of Chicago, in Rudolf Carnap's seminar; he too made early contributions to quantified modal logic. Astonishingly, from 1948 to 1963 Ruth Barcan Marcus, as she became, had no regular affiliation with a major department, and never applied for one. She was a wife and mother, living the life of a housewife and modal logician. However, it was not a life of complete professional isolation: she participated in the life of the greater Chicago philosophical community, and occasionally gave invited lectures or courses. The change of professional name was Alonzo Church's doing, in his capacity as editor of *The Journal of Symbolic Logic*. She had married in 1942, but published under her maiden name until seven or eight years later, when he found out and informed her that future submissions would have to be under her 'legal' name.

Only gradually did the philosophical community realize that she had struck gold, not fool's gold. Of course, gold is fool's fool's gold, but not only fools were deceived: the proper appreciation of her work required a diametric change of philosophical perspective, feeling one's way out of deeply but often tacitly held commitments. It also required a willingness to learn about logic and metaphysics from a woman. Such changes do not happen overnight. Nevertheless, it should have been clear from the beginning that whether or not what she had found was gold, it was at least a mineral of quite unusual quality.

Things improved. Increasing attention was paid to her papers. Philosophical logicians such as Arthur Prior, Saul Kripke and Dagfinn Føllesdal saw their interest and significance. From 1960 onwards, after a gap of seven years, Professor Marcus published a burst of articles in which she reflected on the interpretation of quantified modal logic and answered Quine's criticisms. One of the ideas in them that resonates most with current philosophy of language is that of proper names as mere tags, without descriptive content. This is not Kripke's idea of names as rigid designators, designating the same object with respect to all relevant worlds, for 'rigidified' definite descriptions are rigid designators but still have descriptive content. Rather, it is the idea, later developed by David Kaplan and others, that proper names are directly referential, in the sense that they contribute only their bearer to the propositions expressed by sentences in which they occur. Direct reference entails rigid designation but not vice versa. It was the wildest unorthodoxy when she wrote, and is the purest orthodoxy now.

These papers on modal logic and metaphysics open up and analyse a network of further themes: the nature of extensionality as a principle in semantics; the philosophical groundwork for the necessity of identity; the possibility of a substitutional interpretation of the quantifiers, but also of an objectual interpretation restricted to actually existing concrete individuals, both of which can validate the Barcan formula and its converse; the status of essentialism; the extension of these ideas to properties, sets and other 'collectives'. The discussion is extraordinarily fertile, tersely open-minded and exploratory, as befits the state of the discipline, although still sharply constrained by logic: the emphasis is more on raising questions than on settling them. She is laying out the agenda for a discussion that has been at the heart of philosophy ever since, concerning issues that are as alive now as they were then. Many have contributed substantively to that discussion; there is so much credit to go round that all can afford to be generous over its distribution.

Institutional recognition flooded in too. In 1963, Ruth Barcan Marcus became the founding chair and professor of the Department of Philosophy at the University of Illinois at Chicago, a position of great trust which she held until 1970, building up the department strongly. After three years at Northwestern University, she was then Halleck professor of philosophy at Yale from 1973 to 1992, and subsequently Senior Research Scholar at Yale and Visiting Distinguished Professor at the University of California at Irvine. She was a longstanding Chairman of the National Board of Officers of the American Philosophical Association and President of its Central Division, President of the Association of Symbolic Logic (which she helped achieve financial autonomy) and President of the Institut Internationale de Philosophie and thereafter Honorary President, in addition to extensive service on editorial boards, external review panels and other committees that underpin the collective life of the profession. She held visiting research fellowships at Oxford, Cambridge, Edinburgh, the Stanford Center for Advanced Study in the Behavioral Sciences and the National Humanities Center, and a residency in Bellagio. She was elected a Fellow of the American Academy of Arts and Sciences in 1977. A *festschrift* packed with distinguished authors was published in her honour. She was awarded an Honorary Doctorate by the University of Illinois, the Wilbur Cross Medal by Yale, a medal by the Collège de France, the Machette Prize for contributions to the profession from the Machette Foundation, the American Philosophical Association's Quinn Prize for service to philosophy and philosophers – and, of course, the Lauener Prize itself. Not least amongst those services to the professions was her formidable defence of the highest intellectual standards, of rigour and other core philosophical values, against compromise with fashionable political and cultural pressures. Many professional philosophers, many of them women, testify to her importance to their careers as a mentor. She died on 19th February, 2012, in New Haven.

We do not always expect much activity from great monuments of the profession, but in Ruth's case recognition coincided with a remarkable widening of the range of her creativity. Already in 1966 a pregnant note in *Mind* had helped clarify the interpretation of iterated deontic modalities. Her most-cited paper is one in *The Journal of Philosophy* from 1980 on moral dilemmas and consistency, in which she refuted the popular idea that moral dilemmas involve mutually inconsistent moral principles, and showed that they provide no support for moral anti-realism. She made a powerful case against conceptions of belief that put too much weight on language-use rather than non-linguistic interaction with the external environment, and defended an elegant analogy between belief and knowledge, on which belief requires consistency just as knowledge requires truth. In the history of philosophy, she applied her expertise on modal matters to Spinoza's ontological 'proof' of the existence of God and her ideas on names to the development of Russell's later views on ontology and reference.

The link with Bertrand Russell is no coincidence. A look at the index to Ruth's selected philosophical essays, *Modalities*, shows that he has by far the longest entry of anyone. Like Russell, she uses logic as an essential and creative discipline for philosophy. She is every bit as good as he was at suffering fools gladly. Like Russell, she is willing to try out a variety of ideas with undogmatic experimentation, to follow the argument where it leads, however unpopular the conclusion, while still retaining exactly what he called ‘that feeling for reality, which ought to be preserved even in the most abstract studies’. In reading her work, one has a strong sense that there is truth and falsity in philosophy, just as in other sciences, however hard it is to tell the difference. Sometimes, in sincerely honouring a genuinely distinguished philosopher, one nevertheless feels that in the end all their distinctive ideas will turn out to lie on the false side of the line. So it is a special pleasure to praise Ruth, many of whose main ideas are not just original, and clever, and beautiful, and fascinating, and influential, and way ahead of their time, but actually – I believe – *true*. The award of the Lauener Prize to her must encourage us all to have the courage and patience to carry on the work of analytic philosophy according to the highest standards in our tradition.

Ruth Barcan Marcus

A Philosopher's Calling

Dewey Lecture delivered before the One Hundred Sixth Annual Eastern Division Meeting of The American Philosophical Association in New York, New York, on Tuesday, December 29, 2009

The description provided by the Dewey Foundation reads as follows:

> Each Division would choose a very senior (typically retired) philosopher, someone with a clear tie to that particular division, to give a talk of an autobiographical sort—an intellectual autobiography—with perhaps some account of the way in which he or she was shaped by or shaped the profession, how the profession seems to have changed over the years, etc. The lecturer might reflect on the people and issues that led him or her into philosophy. It is supposed to be a more personal perspective on the life of an important philosopher, a more reflective set of remarks. The lecturer would not be expected to have any particular ties to the work or ideas of John Dewey.

Editor's note: "A Philosopher's Calling"—Ruth Barcan Marcus's John Dewey Lecture delivered before the American Philosophical Association (APA) at New York in December 2009—originally appeared in the *Proceedings and Addresses of the American Philosophical Association* 84, no. 2 (November 2010), pp. 75–92. We thank Ms. Erin C. Shepherd acting on behalf of the American Philosophical Association (which holds the copyright) for granting us permission to republish this autobiographical lecture in the present book.

The prehistory of this gripping, illuminating piece of autobiography goes back to an extemporaneous talk, entitled "Recollections", Ruth Barcan Marcus gave at the 3rd International Lauener Symposium, which took place in her honour in Bern on 30 May 2008. In October 2008, I sent Ruth Marcus, at her request, a DVD of her symposium talk, which she then had transcribed and which most probably served as a basis for the formulation of her fuller autobiographical lecture entitled "A Philosopher's Calling". In addition, Ruth Barcan Marcus planned on reworking the transcript in order to edit her "Recollections" for the present volume. Yet her severe illness discomfitted her. In the course of 2011 it became obvious to me that she would not be in a position to finalize her planned autobiographical piece "Recollections". I therefore suggested to Ruth Marcus that we could instead republish her Dewey Lecture, "A Philosopher's Calling", in our volume, and in October at last, she gladly accepted this proposal with the following words: "I'm sending along a copy of the Dewey Lecture that I wrote for the APA Proceedings; it covers much the same ground as "Recollections"." So the way was found to include in this volume a version of the personal and intellectual recollections which Barcan Marcus gave at the end of the 2008 Lauener Symposium and which we deem to be of great historical value.

I have quoted the description at length so that autobiographical musings are seen by the audience to fall within it. I am surely very senior. Sixty-nine years have passed since I entered graduate school, and jogging my memory has been daunting. It was never my inclination to save papers and records, and in several moves much has gone astray. Much that very likely should have been included has not been retrieved. Additions and corrections are welcome.

I gave no thought to posterity. The philosophical pantheon, even where there is disagreement about membership, will remain small. I once asked a celebrated twentieth-century philosopher well into his seventies and ailing why he tirelessly traveled to give talks despite the toll it took. His reply was, "I do not wish to be forgotten." My thought was that he had, perhaps, gained a year or two in collective philosophical memory.

We are asked to reflect on the issues or persons that led the Dewey lecturer into philosophy. This requires some autobiography. By age three I had somehow taught myself to read by mimicking my older sisters. I recall reading about Chicken Little with the book turned upside down. But then one day the sky was falling right side up. I was reading. It just happened. I also loved numbers and had some computational skills, which were willingly demonstrated. I also had eidetic memory. Some of those abilities gradually faded.

I grew up in a socialist household in the upper reaches of New York City. Our neighborhood was ethnically and religiously mixed, but white and sparsely populated. Over the years it became more densely populated as apartment buildings devoured the lots on which we played. There was a vast neo-Gothic Catholic church at the base of the hill where we lived and a Swedish Lutheran church facing us. The Lutheran church displayed a triangle on which was inscribed, "Seek and ye shall find God is Love." It eluded my understanding. On Sundays there was an influx of worshipers from other parts of the city.

Our sainted figure was Eugene Victor Debs. A bronze bust of Debs graced our living room. The unpictured arch villain was Joseph Stalin. My father's brother remained in what became the Soviet Union, and then disappeared, presumably in Stalin's purge. I recall being at a fundraiser with my father in Madison Square Garden. It was for Norman Thomas, the socialist presidential candidate. Pledges came audibly from the audience and my comment was that Anonymous must be very rich.

At that time socialists and communists were enemies who inhabited parallel universes. They both sang variants of "The International" and carried red flags in May Day parades along different routes laid out by New York's finest. "They" had Young Pioneer and Young Communist League youth groups; "we" had the Red Falcons and the Young People's Socialist League. Which brings me to my first exposure to philosophy: in this case, political philosophy. Red Falcons had various programs, one of which was instruction in Marxist Theory. Marx for tots. I recall being told about the theory of surplus value. An example, presented with illustrations, was a sweatshop where men's trousers were manufactured. Excess inventory accumulated and could not be sold. Workers were made redundant. A depression followed. It was all quite theoretical. The actual depression which we endured came later, but I recall that even at eight (or nine) a causal explanation for a historical event supported by putative arguments and evidence impressed me. Many Marxist slogans such as "Each should produce according to his abilities and receive according to his needs" seemed to me fair, but I noted that it didn't fit observed practice. This set me to explicit thinking about fairness and justice. That particular slogan was more recently echoed in discussions of moral luck.

Nineteen thirty was a catastrophic year. My father died and the Great Depression descended. My mother mourned all her life. We were now a family of four females, which I mention because I believe I had an easier time following an unconventional path than if there had been a strong male presence in the family.

The grade school I attended was confining, rigid, and above all boring. My education was extracurricular. I was a voracious reader and the local library was accommodating. Afternoons, there was also some informal, nonprofit schooling and, weather permitting, there were street games and excursions to local parks. Stickball was the staple. Also red rover, leapfrog, and variations of tag, along with quieter pursuits such as hopscotch, marbles, card games, etc. Girls at that time did not wear trousers. Sometimes my skirts flew and neighbors complained at what seemed to them unseemly behavior, but with little effect. Then, when I was of middle school age, there was a welcome change in my schooling. It was called "progressive education" and was a consequence, I learned much later, of John Dewey's views on proper educational practice. He urged abandonment of the rigid formalities of traditional education. Learning is doing!

A new experimental junior high school was designed for talented students and privately funded by the Ridder family. Admission was to be determined by test and by conferences with teachers, parents, and the candidate for admission. The tests marked me for admission, but I was seen as a "behavior problem," and there was much hesitation as to my admissibility. I had learned enough about plans for the school to realize that it would release me from the bondage of sitting rigidly behind a desk fixed to the floor, forbidden to talk with neighbors and bored out of my skull. I even improvised a prayer to be admitted and it was answered. I was admitted and my expectations were not disappointed. Liberated at last, I was overcome with a desire to learn. Omniscience beckoned. I took to academic work with focus and unbounded enthusiasm. There is not the time to describe my many enthusiasms, but I must mention Euclidean geometry and the concept of rigorous proof.

Two years in junior high, three years in high school, and I was ready for college. At that time it was expected that high-achieving girls from the city would attend Hunter College. Many of them would then make up a pool of prospective teachers for New York City schools. I had no clear future plans, but nonetheless saw that Hunter College was too structured, stifling with uninteresting requirements, and altogether unappealing. It was also single sex. So, after a few days at Hunter, I took myself to Washington Square College of NYU. I enrolled, helped by some NY State Regent Scholarships and the New Deal's funding of student jobs—I assisted the fencing coach at fifty cents per hour. There may have been other sources of financial support which I no longer remember.

Greenwich Village was a ferment. The Philosophy Department was in a building which abutted the location of the infamous Triangle Shirtwaist Factory Fire. The college lounges and cafeteria, the local restaurants and bars, and Washington Square itself were hosts to writers, entertainers, jazz buffs, *Partisan Review* hangers-on, and all manner of political—non-Stalinist—splinter groups (e.g., Musteites, Trotskyites, Lovestoneites, Schactmenites, the Socialist Workers Party, the Workers Party, and others). But the excitement did not distract me from the determination to seek a classical education: mathematics, physics, classics, history, and, of course, philosophy. Since I was an obsessive reader of fiction and criticism, both literary and social, and since a course entitled A Survey of English Literature was a disappointment, literature courses were thereafter excluded from the grand plan but continued to be part of my extracurricular self-education. My dual major was mathematics and philosophy.

The chair of the NYU Philosophy Department was Sidney Hook, one of the former Marxists whose God had failed. He was now a Deweyan pragmatist, and pragmatism dominated the department. Hook was a persuasive lecturer. Under his tutelage the scientific method was now advanced above all others in the pursuit of knowledge. Dialectical materialism had long since been abandoned. His celebrated course was Philosophy of History and Civilization, a subject which has waned. The steps of the experimental (or scientific) method, proceeding from doubt to warranted assertability, were often intoned like a mantra. It was to me a plausible epistemological view. In the department, pragmatism thrived in harmony with logical positivism and its variants, logical empiricism and scientific empiricism. But the positivists, unlike the pragmatists, did not frown on the pursuit of truth or certainty. (Carnap said of philosophy that it is "the logic of science, i.e., the logical analysis of the concepts, propositions, proofs, theories of science,"[1] which meshed well enough with Dewey's pragmatism.) Pragmatists and positivists were later amply represented in the *Encyclopedia of Unified Science* side by side.

My original focus was not to take or defend this or that view. I wanted to understand the important philosophical texts of Plato, Aristotle, the empiricists, Kant, etc. There were courses in the history of ancient, medieval, and modern philosophy. Such study was a facet of my interest in the history of ideas, which was fired in part by an extracurricular reading of Arthur Lovejoy's *Great Chain of Being*.[2] The metaphysical notion of Being was a notion that would not survive for me as empirically meaningful. Fully meaningful or not, it continued to play a vital role in the history of metaphysics. The continuing appeal of such deeply obscure (at least to me) notions was baffling. Similarly, *big* questions like "Why is there something rather than nothing?" or "What is the meaning of life?" baffled me. I was hard put to understand what was being asked.

It wasn't philosophical accounts of meaning which led me to question the sense of such queries. It was my ordinary understanding *via* my ordinary language and my ordinary, commonsense experience. This is usually where I begin my philosophical inquiries. Austin and "ordinary language philosophy" had, and continue to have, a strong appeal.

I have a long memory for popular tunes and lyrics and it was long before my exposure to theories of meaning when I tried to fathom what was being asked by a popular ballad:

> Why was I born? Why am I living?
> What do I get? What am I giving?

What would count as an answer? But, as in the case of my study of literature, my study of those philosophers whom I found deeply interesting but often irrational or obscure (for example, Nietzsche and Schopenhauer) was also extracurricular. I read the texts but took no courses.

At that time, the nineteen thirties, philosophy was of widespread interest, especially political philosophy. At NYU there was a philosophy club, over which I presided, with a university-wide membership, invited speakers, and heated

---

1 Rudolf Carnap. "On the Character of Philosophic Problems." *Philosophy of Science* 1, no. 1 (January 1934): 6.

2 Arthur O. Lovejoy. *The Great Chain of Being: A Study of the History of an Idea* (Cambridge, Mass., 1936).

debate. I recall an occasion when Mortimer Adler spoke and was relentlessly harangued. Hook suggested that, as president, I write an apology.

Interest in philosophy has gradually diminished and will, I believe, continue to wane, or at least its contours will change. Widespread interest among students has declined. There are, to be sure, as there always have been, spin-offs or *quasi* spin-offs—for example, experimental philosophy, cognitive science, neuroethics, bioethics, feminist and transgender philosophy, black philosophy, and so on—but many of the spin-offs do not have settled constituencies. There has also been a revival of interest in traditional metaphysics among some analytic philosophers, partially fueled by an interest in modalities, but philosophy no longer has the widespread undergraduate appeal it once had.

Returning now to NYU's philosophy faculty, there were some who were especially memorable to me but probably unknown to most or all of you. I'll mention two: James Burnham and Albert Hofstadter. Burnham, who with Philip Wheelwright had written a good introductory text, *Introduction to Philosophical Analysis*,[3] was a teacher of extraordinary sensibility, who taught *inter alia* Aquinas and Dante, Thought and Literature of the Renaissance, and Aesthetics. We gained some insight into the medieval and Renaissance world views; we learned how to look at pictures and how to read poetry. When he read to us, it was revelatory. I especially recall his mellifluous readings from Dante and Chaucer. Burnham was an American aristocrat who became a political activist turned Trotskyite after other, radical affiliations. When the German-American Bund goose-stepped its way into Madison Square Garden during the time of the Russian German Pact, we went to protest. Communists were conspicuously absent. Burnham, noticeably styled by Princeton and Oxford, confronted the mounted police—a memorable encounter. But not long thereafter, he returned to his conservative roots. He wrote *The Managerial Revolution*,[4] and *The Machiavellians*,[5] withdrew from academic philosophy, and became a political commentator for *National Review*. Go figure.

There was also Albert Hofstadter, who had collaborated with J. C. C. McKinsey of the Mathematics Department on a paper on the logic of impera-

3 James Burnham and Philip Wheelwright. *Introduction to Philosophical Analysis* (New York: Henry Holt and Co., 1932).

4 James Burnham. *The Managerial Revolution: What Is Happening in the World* (New York: John Day, 1941).

5 James Burnham. *The Machiavellians, Defenders of Freedom* (New York: John Day, 1943).

tives.[6] It was a time when ethics was viewed by some to be grounded in a system of moral imperatives. The paper was an instructive example of how formal methods might be applied to some noncognitive theories. But then Hofstadter too drifted away from logical empiricism. *En route*, he introduced us to Bradley and Bosanquet: my first encounter with Anglo-idealism. Hofstadter, like other faculty whom I admired, was interested in understanding philosophical issues or themes as they evolved over time. They shared the intensity of those interests with their students. Studying philosophy wasn't just an intellectual exercise or an assimilation of texts.

I was also a mathematics major, and, after two logic courses, J. C. C. McKinsey sponsored me for membership in an honors seminar in the Mathematics Department. At the time, there were no suitable texts in English devised for the advanced study of mathematical logic. McKinsey would select a suitable treatise or monograph which could be used. We started with one by D. Hilbert and P. Bernays, *Grundlagen der Mathematik*,[7] which McKinsey translated from the German. He wrote the translation on lined yellow pads and devised exercises. We met once or twice a week for as much as three hours in Bickford's cafeteria, a mainstay of the Village. The fruits of our study were presented to the mathematics honors seminar. It did not occur to me that the attention I received was special, which of course it was. It was, I thought, what college professors did routinely.

I had along the way become interested in C. I. Lewis and modalities, which McKinsey encouraged. Tarski was stranded in New York City at the time, and he and McKinsey collaborated on a study of modal and intuitionistic logic. I worked on proving completeness for Lewis's favorite system, S3, and did not succeed but was not discouraged.

Professors McKinsey and Hook urged me to go on to graduate work. Time and again my mother expressed uneasiness, which was echoed by others: Was it a suitable career? I was eighteen and, as before, wasn't thinking about a career or profession. It was an interest infused with passion. I was advised *not* to apply to Harvard given my interest in modal logic, upon which W. V. Quine, the dominant presence at Harvard, cast a very cold eye. I was advised to apply to Yale, where Professor F. B. Fitch would be more sympathetic. So I applied and was admitted to Yale.

6 Albert Hofstadter and J. C. C. McKinsey. "On the Logic of Imperatives." *Philosophy of Science* 6, no. 4 (October 1939): 446–57.

7 David Hilbert and Paul Bernays. *Grundlagen der Mathematik* (Berlin, Heidelberg, New York: Springer, 1934).

My initial impression of Yale upon arriving in New Haven in 1941 was of a romantic, Gothic fantasy of a university. Despite the difficulties I encountered, the joy at being there was sustained. I sometimes stayed on during breaks so I could inhabit the library, wander about, and pretend possession. But there were problems. (Our "problems" are today's "issues.") Housing for women was separate but not equal. It came with house rules, a curfew, and a housemother. I sought digs elsewhere, where my nocturnal habits were tolerated. We did share a dining room with men, but few women were visible.

In a report in the Yale archives, the graduate dean is reported as saying, "Yale had admitted women to its Graduate School as early as 1892 but had made no provision for their needs as human beings." He also notes,

> My suggestion in 1942 that Yale commemorate in some simple fashion her half century of higher education for women was rejected with scorn by the Secretary of the university who was the *panjandrum* of official ceremonies. His attitude was typical of older Yale College graduates in the faculty and administration. They appeared to be embarrassed by the presence of women in the student body and would have no part in calling attention to Yale's lapse in the matter.[8]

Women were unwelcome in the room of the library where contemporary fiction and criticism were shelved. It was furnished like a comfortable men's club. I protested, and a compromise was reached. Books women wanted would be passed out to be read elsewhere. Also, women could not enter undergraduate classrooms even where they were assisting with grading. There were recurrent discriminatory episodes. I'll mention one graphic incident among many. Yale had a philosophy club open to undergraduate and graduate students. I was elected president but then received a letter from the chair of the department suggesting that I decline. The reasons given were that Yale was predominantly and historically a male institution and that my election may have been a courtesy. Also, the club's executive committee met at Mory's which was closed to women. I did not respond to the letter and did not decline. It was, to me, obviously unreasonable. When the letter was discussed in the graduate dining room, several students said it was imprudent to have revealed it since the chair influences job placement. My response was as before: I wasn't thinking about a job. I assumed the presidency and the executive committee did *not* meet at Mory's (which was not "liberated" until the seventies).

---

8 Edgar S. Furniss. *The Graduate School of Yale: A Brief History* (New Haven: Carl Purington Rollins Printing-Office, 1965), 73.

My work in the department was satisfying. Especially logic with Fitch; Leibniz and Kant's critiques with Ernst Cassirer; philosophy of science with Cassirer, Northrup, and Margenau (of the Physics Department); and wonderful discussions with Charles Stevenson and Monroe Beardsley. Cassirer was a stellar teacher and a polymath who could convert positivists into Kantian idealists.

I had a copy of the second edition of *Principia Mathematica*, which was purchased out of a $100 book prize awarded to a first year student and which I began to study systematically. I then went on to study Russell's *Principles* and *Introduction to Mathematical Philosophy*. I discovered "On Denoting" while looking at old copies of *Mind* in the Sterling Library. Russell remains a primary philosophical influence.

We were free to take whatever courses or tutorials were available in and out of the department. I audited mathematics courses taught by some of the strong mathematics faculty and enrolled in a six day a week, year-long course in theoretical physics required of first-year graduate physics students: five lectures with a test on Saturday. The major Philosophy Department requirement was the Preliminary Examination; a five- (or was it six?) day series of examinations in the "fields" of philosophy. It was a marathon. I've watched, with some rue, the requirement disappear.

With Fitch's approval, I began to work on a dissertation proposal. There was an entrenched belief that there were insurmountable difficulties of interpretation with extending modal propositional logic to include quantification. I did not see such insurmountable difficulties. My plan was to develop rigorous axiomatic extensions of some of Lewis's systems to include quantification. Problems of interpretation could then be discussed relative to the framework. Meanwhile, as I worked on the prospectus, the Second World War was escalating, and science departments were increasingly shorthanded as faculty departed for war-related research. I was persuaded to take on grading for the graduate theoretical physics course, for which I received $550. I then learned that the amount had to be deducted from my miniscule fellowship. (Yale largesse.) Grading that course was an ordeal. The language of physics was not my native tongue.

There was an influx of reserve officer training candidates, and my husband, who was a physics graduate student, was appointed "Instructor." Yale wheels began to grind. Since I was now a faculty wife, however accidental, a gorgeous bouquet was delivered. The provost's wife called and left an engraved card. I had cards engraved and returned the call. I still have the remaining cards.

By the fall of 1943 my dissertation prospectus, *A Strict Functional Calculus*, was approved. In 1944 we moved to the Washington, D.C., area, where my husband was now engaged in military research at the Johns Hopkins Applied

Physics Laboratory. I completed the dissertation *in absentia*, using the Library of Congress as needed. The library was in disarray, and when a request for a book or journal like the *Journal of Symbolic Logic* was delayed, the explanation would be that members of Congress had special borrowing privileges.

I mailed sections to Fitch for perusal. He finally said I had done enough and suggested that I immediately submit an initial part on first order modal quantification theory for publication in the *Journal of Symbolic Logic*, which I did. It was accepted without delay by the then editor Alonzo Church, published in *JSL* 11[9] and reviewed the same year (1946).[10]

With the end of the war I returned to New Haven. My dissertation was actually ready for submission in 1945, but it was submitted for the 1946 deadline. Church accepted two other papers for *JSL*: one on the deduction theorem in first order modal quantification theory[11] and one on second order modal quantification with identity, in which the necessity of identity was proved.[12] That resulted in a stir since, at the time, the existence of contingent identities was a received view.

Church, noting my Russellian dispositions, then asked me to review a paper of Arthur Smullyan, who defended an early Russellian solution to substitution puzzles in modal contexts.[13] I defended Smullyan in turn.[14] It was about the time of the review request (my recollection is fuzzy here) when Church informed me testily that he had learned I was married and must heretofore use my "legal" name on papers submitted to the *Journal of Symbolic Logic*. I acquiesced.

A word about publication: I am not driven to publish. Papers are submitted where I think I have a useful account of or solution to a clear question of logical or philosophical interest. The questions usually originate in some common sense observations, couched in our common, ordinary language. It is therefore disappointing when blatant errors about what I have done occur and persist. I

---

9 Ruth C. Barcan. "A Functional Calculus of First Order Based on Strict Implication." *The Journal of Symbolic Logic* 11, no. 1 (March 1946): 1–16.

10 W. V. Quine. Review of "A Functional Calculus of First Order Based on Strict Implication," by Ruth C. Barcan. *The Journal of Symbolic Logic* 11, no. 3 (September 1946): 96–97.

11 Ruth C. Barcan. "The Deduction Theorem in a Functional Calculus of First Order Based on Strict Implication." *The Journal of Symbolic Logic* 11, no. 4 (December 1946): 115–18.

12 Ruth C. Barcan. "The Identity of Individuals in a Strict Functional Calculus of Second Order." *The Journal of Symbolic Logic* 12, no. 1 (March 1947): 12–15.

13 Arthur Francis Smullyan. "Modality and Description." *The Journal of Symbolic Logic* 13, no. 1 (March 1948): 31–37.

14 Ruth C. Barcan. Review of "Modality and Description," by Arthur Francis Smullyan. *The Journal of Symbolic Logic* 13, no. 3 (September 1948): 149–50.

mean literal errors—not disagreements about interpretation. In the case of the paper on identity, for example, a major result and its import were missed by the reviewer. I expected the error would be noticed and corrected, but after *eleven years* of expectation, during which the error had been carried along by others in the literature, I wrote to the reviewer, who then informed Church: "A grave and puzzling error in my review XII 95(4) of Miss Barcan has just come to my attention. It is ancient history, but still I'd feel relieved if you could see your way to publishing a signed correction." A correction was published in *JSL* XXIII. Misreadings and neglect of some later work continued, but not uniformly. Some misreadings and omissions were corrected, some escalated into controversies, and some results were ignored. My keen disappointment was that my romantic notions about the self correcting feature of research within a scholarly community were not a given. There remain lengthy bibliographies and historical accounts of intensional and modal logic as well as interpretations of modalities where reference to my work is absent, but that is gradually being corrected.

Returning now to my recollections, we were back in New Haven where my husband began a dissertation in low temperature physics. In the absence of local teaching opportunities I took a research position at the Yale Institute for Human Relations. The Institute, funded by the Rockefeller Foundation, was dominated by Freudians, and an orthodox analysis was required for some PhD candidates in psychology and social anthropology. Yale was also a center for the study of behaviorism, under Clark Hull. I was recruited to participate in two areas of research which were being pursued. First, the attempt to verify some psychoanalytic hypotheses cross-culturally, such as relating childhood trauma to irrational adult explanations of illness in cultures without scientific medicine. Several of us participated. There was secondly, turned over to me, a project of synthesizing psychoanalytic and behavioral theory. Neither project was a success. Thinned out results were finally published in a book by J. Whiting and I. Child, *Child Training and Personality*, in which I am described as a "contributor and critic of concepts."[15] But I welcomed the opportunity to study psychology and social anthropology.

With PhDs granted, we were off to Chicago with post doctoral fellowships, mine from the American Association of University Women. It was for me an opportunity to study with Carnap. Carnap was also concerned with modalities and quantification, but we were in strong disagreement. In his modal language individual terms had dual reference; in the scope of a modal operator the

15 J.W.M. Whiting and I. Child. *Child Training and Personality: A Cross-Cultural Study* (New York: Yale University Press, 1953).

referent was an intension (an individual concept)—something that I was at pains to resist. I welcomed an opportunity to discuss our differences.

A day after arrival, I inquired at the University of Chicago about making suitable arrangements. The chair of the department gave me Carnap's home address. It seems he met students at home and rarely frequented the department. The atmosphere was strained. Carnap's autobiographical remarks, published in the Schilpp volume on Carnap, are revealing. He says of a departmental discussion, "I had the weird feeling that I was sitting among a group of medieval learned men...and would dream that one of my colleagues raised the famous question of how many angels could dance on the point of a needle."[16] Such a remark might suggest that Carnap lacked appreciation for the history of philosophy, which was clearly false. He did believe that analytic philosophy had made important advances worthy of consideration but which were not regarded as such by many of his colleagues. Some junior faculty warned me, in jest, that if the Chicago department invited me to give a talk I might be asked whether my views were to be found in Nicholas of Cusa.

I joined some young scholars who met regularly in Carnap's flat. They gathered around him. (He was prone due to a back ailment.) Some of the graduate students, for example, Dick Jeffrey, believed that completing a dissertation under Carnap would be thwarted by the department. Jeffrey transferred to Princeton. I was rapidly learning that departments were not the peaceful communities of scholars I had fantasized.

After two years, we moved north to Evanston, Illinois, where my husband joined the Northwestern University Physics Department. Northwestern at the time had a nepotism rule, which excluded regular appointment of spouses. In any case, I did not seek a regular position. My publications were well received. I had some courtesy privileges and was invited to participate in the colloquia of the department and to teach an occasional course as a "Visiting Professor." In 1952 I was awarded a Guggenheim Fellowship. Between 1959 and 1962 I continued to think about modalities and intensional languages informally and formally as well as interpretively. I also taught part time at Roosevelt College (named after FDR) in Chicago. It was originally a YMCA urban school, but, when faced with racist demands, the administration and most faculty agreed to secede from the Y. It then occupied one of the architectural marvels designed by Louis Sullivan: the Auditorium Hotel overlooking Lake Michigan.

---

16 Paul Arthur Schilpp. *The Philosophy of Rudolf Carnap*, 1st ed., The Library of Living Philosophers v. 11 (La Salle, Ill: Open Court, 1963).

There was, astonishingly, *no* state-supported four-year college in Chicago. A comprehensive university was projected, but political wrangling delayed it for many years. It was a matter of competing for many millions of dollars and acres of Chicago real estate which impinged on ethnic neighborhoods. A temporary junior college of the University of Illinois system had been established on Navy Pier, which was supposed to transition into the college, but the project languished.

My research job at Yale had stimulated an interest in psychoanalysis, and I was accepted as a "control" at the Chicago Psychoanalytic Institute, meeting four times a week with an analyst-in-training. Interesting as it was, my research at Yale and my orthodox analysis were consistent with the later conclusions of Grunbaum et al: psychoanalysis as theory or clinical practice would not bear scientific scrutiny.

I was in communication with philosophers who were based in the area and those who visited. There was Leonard Linsky in the Chicago department with whom I had many fruitful and continuing discussions. He later included my paper "Extensionality" in the Oxford readings, *Reference and Modality*.[17] During that period, I met David Kaplan when he visited Chicago. He has remained a very insightful critic and a fast and loyal friend. Arthur Prior gave a seminar at the University of Chicago in 1962 which I attended. I believe it was Prior who coined the term "Barcan Formula" for an axiom about the mingling of modal operators and quantifiers, about which debate continues.

In 1961, the Boston Colloquium in Philosophy of Science invited me to present the paper "Modalities and Intensional Languages." Quine's negative views had been expressed immediately on the heels of my publications, in his *JSL* reviews of 1946 and 1947 as well as an article, "The Problem of Interpreting Modal Logic," also in the 1947 volume of *JSL*, where he considers "the systems of Miss Barcan."[18] It was as if such efforts needed to be nipped in the bud. My address was published in *Synthese* in 1961 and presented in February 1962 with

17 Leonard Linsky. *Reference and Modality*, Oxford Readings in Philosophy (London: Oxford University Press, 1971).

18 Quine, "Review."

W. V. Quine, Review of "The Identity of Individuals in a Strict Functional Calculus of Second Order," by Ruth C. Barcan, *The Journal of Symbolic Logic* 12, no. 3 (September 1947): 95–96 (Corrected in JSL 1958).

W. V. Quine, "The Problem of Interpreting Modal Logic," *The Journal of Symbolic Logic* 12, no. 2 (June 1947): 43–48.

comments by Quine.[19] The entire proceedings was published in *Boston Studies in Philosophy of Science* along with some much edited further discussion by Quine, Føllesdal, Kripke, and McCarthy.[20] I was apprehensive about the colloquium. My name is Ruth and I was in alien corn: the Harvard Faculty Club.

There is, on my account, no inflated metaphysics of possible worlds, except as a *façon de parler*. Possibility is about the way the actual world might be. Three axiomatic extensions of Lewis's systems with quantifiers are considered. The domain of interpretation consists of *actual* individuals. A theory of direct reference is proposed for proper names of individuals. They do not change reference in the scope of modal operators as in Carnap. Descriptions do not refer directly—they describe. They function like predicates or attributes. There are no *possibilia* (possible individuals). Possibility is about properties actual things might or do have. A modal language will accommodate talk of essential attributes, i.e., necessary but not logically necessary attributes, but may be consistent with the falsehood of all essentialist claims—a conclusion I intuited and Terry Parsons later proved in "Essentialism and Quantified Modal Logic."[21] But I saw nothing amiss in what is genuine "Aristotelian essentialism," which Quine characterized as "invidious." Aristotelian essentialism is about essential properties—not about individual essences. Candidates for essential attributes were, as I understood it, physically necessary properties: those covered by physical or more broadly empirical law. They are not metaphysical. I do tend to call "metaphysical" some high level *physical* laws such as the one Einstein proposes in the special theory of relativity: Laws of physics are the same when determined relative to one inertial system as when determined relative to any other.

An aside: The question which baffled Russell was how an ordinary proper name could retain its reference over time, where there is no direct access to the object named. My thought was that Russell should not have abandoned what seemed to be his earlier acceptance of direct reference for proper names (e.g., "Scott"), although how that could be (i.e., direct reference over time) was not yet explained. Russell later identified "logically" proper names with a pointing,

19 Ruth Barcan Marcus. "Modalities and Intensional Languages." *Synthese* 13, no. 4 (December 1961): 303–322; W. V. Quine. "Reply to Professor Marcus." *Synthese* 13, no. 4 (December 1961): 323–330.

20 Marx Wartofsky, ed. *Proceedings of the Boston Colloquium for the Philosophy of Science, 1961–62*, vol. 1 of *Boston Studies in the Philosophy of Science* (Dordrecht: D. Reidel, 1963), 77–112.

21 Terence Parsons. "Essentialism and Quantified Modal Logic." *The Philosophical Review* 78, no. 1 (January 1969): 35–52.

accompanying "this" and "that," and interpreted ordinary proper names as disguised descriptions: a backward move. The first satisfactory answer to Russell's question was, to the extent that I can determine, that of Geach several years later in "The Perils of Pauline," where he sets out the historical chain theory:

> For the use of word as a proper name there must in the first instance be someone acquainted with the object named. But language is an institution ... and the use of a name for a given object ... like other features of language, can be handed down from one generation to another, ... Plato knew Socrates, and Aristotle knew Plato, and Theophrastus knew Aristotle and so on in apostolic succession to our own time; that is why we can legitimately use 'Socrates' as a name the way we do.[22]

Prior to Geach, I had dug my heels in. It was commonly claimed that lexicographers said of proper names that they had no "lexical meaning," but they, too, to the extent I could determine, gave no answer to Russell's question. Nor did I. I was persuaded to the point of stubbornness that one would surely be forthcoming. It was, in the historical chain theory of Geach, later alternatively (but less perspicuously) described as a causal theory of names.[23] Quine, in his response to my talk, dismissed the direct reference view as a "red herring."

In August 1962, following the February Boston Colloquium, through the good offices of Von Wright and Hintikka, an international "Colloquium on Modal and Many Valued Logics" was mounted in Helsinki, sponsored by the International Union of History and Philosophy of Science. A proceedings was published in *Acta Philosophica Fennica* (1963), which included papers by Aqvist, Geach, Hallden, Hintikka, Kripke, Lemmon, Montague, Prior, Rasiowa, Rescher, and Smiley, all of which would be of interest to those concerned with modalities.[24] I presented a paper on sets and attributes (later revised for improved vocabulary). The colloquium was exciting and productive. A cottage industry on modalities was launched and continues.

Meanwhile, the University of Illinois in Chicago had settled on a location and an architecturally *avant garde* campus rose on what was an embattled ethnic Greek neighborhood. A search was launched in 1963 for departmental chairs, and I was approached. It seems I was not a unanimous choice of the search committee. The protest was that there never had been a woman chair at

---

22 Peter Geach. "The Perils of Pauline." *Review of Metaphysics* 23 (December 1, 1969): 288–89.
23 See Lecture II of Saul Kripke, "Naming and Necessity," in *Semantics of Natural Language*, ed. D. Davidson and G. Harman (Dordrecht: Reidel, 1972).
24 *Acta Philosophica Fennica*, vol. 16, 1963.

the University of Illinois! But it happened, and I embarked on one of my careers as a proper "professional." In a short time we had a strong and congenial faculty, including (in senior and mid level positions) George Dickie, Arnold Levison, Terence Parsons, Bryan Skyrms, William Tait, Irving Thalberg, and Paul Ziff. Also, more junior, Fred Feldman, Paul Teller, and Rudolph Grewe. Ian Hacking came one term a year. Two instructors with *de facto* tenure as well as a young, able, feminist "continental" philosopher, Sandra Bartky, from Navy Pier were invited to join. There were other able adjunct appointments, such as Marcia Eaton. Money was raised to improve library holdings. Those holdings were enhanced by our purchase of Cassirer's personal library. Our plans for a graduate program were closely scrutinized by external examiners and approved. Initially, graduate students from the University of Chicago were recruited as graduate assistants. We decided not to employ Straussian graduate students from the University of Chicago Committee on Social Thought, who taught an esoteric code for deciphering philosophy texts. But soon enough we had our own graduate students. Our first two PhDs were Nancy Cartwright and Vivian Weil. The Society for Women in Philosophy was organized under Sandra Bartky's initiative. It became, and remains, a strong department.

Another of my parallel "careers" was participation in the work of our professional associations. In 1961, Charles Stevenson, then at Michigan, requested my services as secretary of the Western (now Central) Division of the APA. I was prompted to take it on by an awareness of the extent to which practices of professional associations impinged on their constituencies. Employment practices needed reform. Other demands on professional associations, such as evaluations of programs, defending the rights of its members and the like, needed to be scrutinized. I continued serving the APA for fifteen years, some of it in the Central Division, including the office of president, and concluded with two terms, totaling six years, as chairman of the National Board of Officers in 1982. The Association was transformed on the national level into a constitutionally defined professional association. The old boy system of recruitment was transformed and reforms initiated. Standing committees on a range of issues concerning philosophers and philosophy were established.

Between 1963 and 1986 I also held various offices in the Association of Symbolic Logic, serving as vice president 1980–83 and president 1983–86.

Between 1960 and 1980, other papers I wrote which were of significant interest were "Dispensing with Possibilia," "Iterated Deontic Modalities," "Essentialism in Modal Logic," "Quantification and Ontology," "Nominalism and the Substitutional Quantifier," and "Essential Attribution." A paper titled "Moral Dilemmas and Consistency" had a surprising impact and is included in many collections on ethics. It argues *inter alia* that on logical accounts of consistency, moral dilemmas need not be a mark of inconsistency of a moral code.[25]

In 1970, Northwestern abandoned its nepotism rule and invited me to join the department. It was decidedly a "continental" department with the very able Sam Todes and William Earle as colleagues. Hilary Putnam came from UCLA on his first appointment, but left after a year. Shortly after I joined, I received calls from Yale. There had for many years been steady departures of faculty from Yale: in 1967 Wilfrid Sellars, along with Alan Anderson, Nuel Belnap, and Jerry Schneewind, and followed by Rich Thomason, decamped for Pittsburgh, where they became the core of a continuing strong program.

Kingman Brewster, the then Yale president, urged me to come and help rebuild the department. In 1972 I finally agreed and arrived in 1973. Memories in New Haven were still vivid of the protest against the Vietnam War and the shutdown of the Yale campus due to the trial of black activists.

I was one of three tenured faculty women. Although upper division women had been admitted as undergraduates, the admission of lower division women and the equalization of numbers and facilities were still being debated. Faculty meetings were tense. Yale had been long self-described as educating "Leaders of Men." There was protracted discussion of the need for hair dryers and separate bathrooms and the attendant costs. Over time, women came to be admitted on a more equal basis. Women's Studies and various programs and facilities for women were established. Mory's was liberated.

The Yale Philosophy Department was unsettled. Stephen Korner remained until retirement. Robert Fogelin left after fourteen years. Sarah Broadie came and then left for Princeton. Bob Adams came, served as chair, and then left for Oxford. It was only partly the result of a continental divide. Despite many handsome competing offers, I stubbornly stayed.

The dominant presences in the humanities at Yale were de Man, Derrida, and deconstruction. Derrida was sponsored by de Man, the reigning guru of the Comparative Literature Department. The *New York Times* ran a magazine story

---

25 Ruth Barcan Marcus, "Dispensing with Possibilia," *Proceedings and Addresses of the American Philosophical Association* 49 (1975): 39–51; Ruth Barcan Marcus, "Iterated Deontic Modalities," *Mind* 75, no. 300, New Series (October 1966): 580–82; Ruth Barcan Marcus, "Essentialism in Modal Logic," *Noûs* 1, no. 1 (March 1967): 91–96; Ruth Barcan Marcus, "Quantification and Ontology," *Noûs* 6, no. 3 (September 1972): 240–50; Ruth Barcan Marcus, "Nominalism and the Substitutional Quantifier," *Monist: An International Quarterly Journal of General Philosophical Inquiry* 61 (July 1, 1978): 351–62; Ruth Barcan Marcus, "Essential Attribution," *The Journal of Philosophy* 68, no. 7 (April 8, 1971): 187–202; Ruth Barcan Marcus, "Moral Dilemmas and Consistency," *Journal of Philosophy* 77 (March 1, 1980): 121–36.

pro and con, "The Tyranny of the Yale Critics," which included a sample of deconstructionist interpretation. A popular critic wrote, in a letter to the editor, that he thought the sample was the answer to the previous week's double-crostic.[26]

Derrida was a recurrent visitor in Comparative Literature, but Philosophy resisted a joint appointment. The expressed negative views of John Searle and me earned us a footnote in Derrida's book *Limited Inc.* He describes us as "members of an academic Interpol."[27] With the disclosure that de Man had written extensively for a Belgium newspaper sympathetic to Nazism and had invented a fanciful persona in the United States as a putative *émigré* from the Belgium underground, deconstruction left Yale.

Despite the unevenness in the philosophy department, it was my professional home. However, I moved about: The Stanford Center for Advanced Study in the Behavioral Sciences 78–79; Fellow of Wolfson College, Oxford, Trinity terms 85–86; Clare Hall, Cambridge, Trinity term 1988, where I was made a permanent member of the common room. I enjoyed a stimulating term at UCLA. There were two summers as a guest of the Rockefeller Foundation's residence for scholars on Lake Como. In 1973 I was elected to the Institut International de Philosophie, proceeding through various offices, including president 1989–92 and, presently, president honoraire. I attended its far flung meetings. I was, for several years, a member of the steering committee of the International Federation of Philosophical Societies. In all of these excursions I encountered philosophers with whom I could profitably exchange views. I also encountered many ideological differences and disagreements and the degree to which political differences impinged on the scholarly profession. I'll cite two examples, out of many. I was program chairman of the Seventh International Congress of Logic, Methodology, and Philosophy of Science, held in Salzburg, Austria, from July 11 to July 16, 1983. G. E. Minc, of the Soviet Union, was invited by the section on model theory and proof theory to submit a paper, which he did, on applications of proof theoretic transformation. Dr. Minc had previously requested emigration from the Soviet Union and had been consequently relieved of his research position. We were informed that since he was unemployed he was denied exit to attend the conference. The bylaws of the International Union require that all

---

26 Colin Campbell, "The Tyranny of the Yale Critics," *New York Times* (1857–*Current file*), February 9, 1986. Available at http://www.proquest.com. Accessed February 12, 2010; Andrew A. Rooney, "Letter to the Editor 2 – No Title," *New York Times* (1857–*Current file*), March 16, 1986. Available at http://www.proquest.com. Accessed February 12, 2010.

27 Jacques Derrida. *Limited, Inc.* (Evanston: Northwestern UP, 1988).

scholars have free mobility to and from conferences, but my efforts at gaining him a visa from the Soviet authorities were not successful. Professor Feferman agreed to read Minc's paper, but at a meeting of the program committee, Russian delegates said there was no rule which permitted someone's paper to be read. It was heatedly debated, but we pointed out there was ample precedent for this being done, and finally Feferman did deliver Minc's paper. After the meeting, several of us tried to help Minc in his efforts to emigrate. He finally succeeded and is presently a professor in the Stanford Mathematics Department. Nor is it the case that the United States was free from such decisions. Iris Murdoch told me that in 1946 she was denied a visa to this country, and subsequently refused invitations to visit the United States, excepting one occasion on which she was invited to lecture and was granted a restricted visa confined to the city in which she was to lecture. One case among many in the United States during the 1940s and subsequent years.

A high point was an invitation to lecture at the Collège de France in 1986—an interesting institution. The Collège was originally established by Francis I with the purpose of bringing France up to a higher standard of research, which had been thwarted at the Sorbonne. The lectures are advertised throughout Paris and many of the audience come in off the street. There were stern ladies who knit like Madame Defarge. I apologized for having to lecture in English but was excused since, as I was told, "The last American who lectured at the Collège only *thought* he was speaking French."

I served on panels of the National Science Foundation, the National Endowment for the Humanities, the Rockefeller Foundation, the Fulbright Committee, and the Committee for Philosophy of the Educational Testing Service. We recommended that the Educational Testing Service abandon the philosophy test. There were frequent strange imbalances in that test: one year, for example, an inordinate number of questions concerned Paul Tillich, whereas major figures were neglected.

From 1979 onwards I served as a visitor and evaluator of programs at many universities, including Princeton, MIT, Columbia, Caltech, Duke, University of Massachusetts Amherst, and some University of California campuses. I also gave invited lectures and participated in conferences in the United States and abroad. The APA mounted a session on my work. Many further details of my "career" are here omitted, lest this memoir read like a laundry list. Among papers written after 1980, there are three on belief and believing which are critical of the dominant language-centered view of belief and suggest some revisionary proposals. Of some interest is a study of Quine's animadversions on modalities and one on Russell's views of particularity.

It is true that the Yale department was fragile, but the students were a joy. I taught introductory philosophy in a special program for freshman, undergraduate survey courses in ethics, two or three levels of logic, and diverse advanced courses. I cherish the letters of thanks from students over the years. It is they who deserve thanks for the pleasure and stimulation they afforded. Some of the undergraduates I remember went on to careers in philosophy. Among those I recall are Tim Maudlin, Jessica Moss, Adina Roskies, Chris Smeenk, and Susan Wolf. Some of my graduate students are distinguished professors: Walter Sinnott-Armstrong, Nick Asher, Frank Farrell, Don Garrett, and Diana Raffman. Sinnott-Armstrong, in collaboration with Raffman and Asher, edited a *festschrift*, *Modality, Morality and Belief.*[28] A later *festschrift* was published in *Dialectica* (1999), edited by Henri Lauener.

Philosophers from whose views I have profited (although we are often in disagreement) are, in addition to those already mentioned, Nuel Belnap, Paul Benacerraf, Max Cresswell, Kit Fine, Robert Fogelin, Pat Greenspan, Henri Lauener, Isaac Levi, Charles Parsons, Terry Parsons, Ori Simchen, Ernest Sosa, Bob Stalnaker, Judy Thomson, and David Wiggins. But I am essentially a loner. One of the changes in academic style in recent years is the distribution of papers by an author for comment by large, sometimes astonishingly large, numbers of contemporaries, which is then noted in the acknowledgments. That was not my style. There was often no point, in any case, since I characteristically defended positions contrary to received views, if there were received views. I resist the pigeonholing philosophical taxonomy such as "materialist," "idealist," "realist," "dualist," "functionalist," "internalist," "physicalist," etc., etc. I find myself in support of Pen Maddy's low key "naturalistic views" as given in *Second Philosophy: A Naturalistic Method.*[29] And I do use the term, as in the paper "The Anti-Naturalism of Some Language Centered Accounts of Belief."[30]

Mandatory retirement came in 1992. A collection of essays, *Modalities*, was published by Oxford.[31] The many critical reviews were favorable beyond expectation. I continued to teach an occasional course at Yale, and I visited the University of California at Irvine for some consecutive years, one term a year, with

---

28 Walter Sinnott-Armstrong, Diana Raffman, and Nicholas Asher, eds. *Modality, Morality and Belief: Essays in Honor of Ruth Barcan Marcus* (Cambridge University Press, 1995).

29 Penelope Maddy. *Second Philosophy: A Naturalistic Method* (Oxford University Press, USA, 2007).

30 Ruth Barcan Marcus. "The Anti-Naturalism of Some Language Centered Accounts of Belief." *Dialectica* 49, no. 2–4 (1995): 113–130.

31 Ruth Barcan Marcus. *Modalities: Philosophical Essays* (Oxford University Press, USA, 1993).

pleasure and profit. After 1998, I no longer traveled abroad with the exception of a trip to Bern in May 2008 for an International Symposium on Analytical Philosophy in my honor, sponsored by the Lauener Foundation. Henri Lauener was a colleague and member of the International Institute of Philosophy who edited a *festschrift* published in 1999.[32] There were challenging papers, to which I have still to respond. There was music, an address by Professor Essler, who is president of the foundation, and a head turning "laudatio" by Tim Williamson.

Before closing, I want to express my gratification at the present benign state of the Yale department after many years during which I began to despair. It is Michael Della Roca, a fine philosopher and unparalleled administrator, who has chaired for over six years and steered the Department to its present level of distinction—a Sisyphean task. My thanks finally to the APA for inviting me to give this lecture and my thanks to all of you for your patience.

32 Henri Lauener, ed. *Dialectica* 53, no. 3–4 (1999).

Dagfinn Føllesdal

Ruth Marcus, Modal Logic and Rigid Reference

Abstract

This paper surveys Ruth Marcus' many contributions to modal logic and its interpretation, starting with her pioneer work on quantified modal logic and ending with the controversies concerning the origin of the idea of rigid reference and other basic ideas in the so-called "New theory of reference." Her contributions are discussed with close attention to who gave credit to whom.

Nobody is referred to more often than Ruth in my 1961 dissertation. Her work in modal logic is thoroughly discussed, starting with her four contributions to the JSL in 1946–1948. I first met Ruth in January 1962, when she visited Harvard to give her lecture "Modalities and Intensional Languages." This led to a friendship that has now lasted for 46 years. Looking out over all this youth, I am confident that I must be the one here who has known Ruth the longest.

I later discovered that I was not the first Norwegian in Ruth's life. She writes in the acknowledgments of her book *Modalities* that after McKinsey who taught her as an undergraduate, two of her important teachers were Einar Hille and Øystein Ore, whose mathematics lectures she audited at Yale. Mathematics and physics, she writes, "helped shape my *philosophical* thought."[1] Both Hille and Ore were Norwegians. Ore, by the way, was no minor figure in mathematics at that time. Eric Temple Bell, in his history of mathematics, regarded him as one of the major mathematicians of the twentieth century, much on the basis of his creation of lattice theory in the early thirties, which at that time was regarded by many, including Bell, as an innovation on a par with group theory.

However, now from Ruth's early life to her *work*.

As I mentioned, Ruth's contributions to the JSL started in 1946. It is unbelievable, when you see her sitting here with us one should think she must have been a child prodigy. But at that time she already had her PhD – very young, but not a child. And her work was important. In the first of the four articles I men-

Stanford University and CSMN, University of Oslo

1 Barcan Marcus 1993, p. x. My emphasis.

tioned, in March 1946, she developed a *quantified* modal logic. She was the first ever to do this, until then all modal logic had been propositional. Carnap published his version of quantified modal logic a few months later, but Ruth was first.

Ruth's quantified modal logic opened up a whole new field of research. First and foremost, it gave a scope and flexibility to modal logic that had been lacking in propositional modal logic; without quantifiers one cannot even formulate some of the issues that modal logic is designed to solve. As Carnap puts it in *Meaning and Necessity* (1947):

> Any system of modal logic without quantification is of interest only as a basis for a wider system including quantification. If such a wider system were found to be impossible, logicians would probably abandon modal logic entirely.[2]

Hintikka has made the need for a quantified modal logic evident by showing how in the case of deontic logic introduction of quantifiers must be regarded not only as a means of making current systems more comprehensive, but as, [I quote] "indispensable for any satisfactory analysis of the notions with which *every* system of deontic logic is likely to be concerned."[3] By showing that central notions in ordinary language like those of obligation, forbiddance, permission, and commitment, call for an analysis in terms of quantifiers, Hintikka has, it seems, not only made a good case for the *need* for a quantified modal logic, but also, unless our ordinary reasoning and discourse in this area is extraordinarily muddled and confused, for the *possibility* of one. Added support for the view that quantified modal logic is possible is rendered by the fact that actual systems of quantified modal logic have been constructed and claimed to work. Ruth's work is a major contribution here.

In addition to its crucial importance for the applications of the modalities, quantified modal logic also opens up a new field of formal research. The interplay between quantifiers and modal operators and the structure of this new field was studied intensively in the years that followed Ruth's 1946 paper. Ruth made numerous important contributions to this field. In my dissertation I surveyed this developing field and pointed out that if one adds to the field of quantified modal logic the new ideas concerning the semantics of iterated modalities that had been proposed and developed in 1957 by Stig Kanger in his

2 Carnap 1947, 1956. Page 196 of the 1956 edition.
3 Hintikka 1957b, p 3.

dissertation *Provability in Logic* and Hintikka in two articles from that same year, one gets new insights and new questions.

In my dissertation I presented and explained these ideas, which had not been presented in detail before. Of course, I claim no originality for these ideas, but give all the credit to Kanger, in particular, and also to Hintikka. These ideas, the chief of which is to talk not simply of possible worlds but of worlds as being possible relative to one another, are now usually called Kripke semantics, although Kripke presented these ideas as late as in 1962 at a colloquium in Helsinki. Kripke clearly contributed a lot to giving a clear and pedagogical presentation of these ideas and developing them further. However, I think Kanger deserves much more credit than he has received and I have therefore proposed to call this approach "Kanger-Kripke semantics."[4]

However, enough about priorities at this point. In my dissertation I showed that Ruth's results concerning iterated modalities got a particularly natural interpretation in a system where the relation between possible worlds is symmetrical as it is in the so-called "Brouwer system." For example, the Barcan formula, '$\forall x \Box Fx \supset \Box \forall x\, Fx$', about which Tim Williamson said we would hear more today, and its converse, get very straightforward proofs in this system. Thus the semantic considerations guide us towards finding simple proofs; the Barcan formula gets a shorter and simpler proof in Brouwer's system than it gets in the stronger, traditional S5. In my dissertation I proved, in simple ways, a number of theorems concerning mixtures of quantifiers and modal operators that become salient in this kind of semantics.

These are, however minutia, unimportant byproducts which I mention only because they came out of Ruth's pioneering work 15 years earlier, to which I refer at least 40–50 times in my dissertation. Now to the philosophical issues.

The survey of what had been done in modal logic before 1961, in which Ruth's work played a very important part, was just providing a background for my main aim in the thesis, which was to examine critically my thesis advisor Quine's arguments against modal logic. In spite of my very high respect for Quine – there is no philosopher I regard more highly – I found that there was something wrong with his arguments against modal logic, and I set out to find out what it was. Modal logicians had mostly shrugged his arguments off by saying: "We *do* modal logic and prove theorems – what is the problem?" One of the impressive features of Quine – whose centennial we are celebrating these days – was his *problem sensitivity*. His search for clarity was so astute that he

4 Føllesdal 1994.

spotted obscurity and fundamental difficulties where others thought there was smooth sailing. So with analyticity and many other problems and so with modality.

Quine had been building up his case against the modalities for decades. What began already in his dissertation (1932) as misgivings about meaning, moved on to the modalities in the early forties, since there one could see some of these problems in a clearer context.

Gradually, Quine argued that clarity of ontology requires *identity* to be *universally substitutive*, not just substitutive in extensional contexts – as many modal logicians claimed, but not Ruth. She proved already in one of her first papers that *identity is necessary* and hence universally substitutive. Quine in his review of this paper[5] pointed out that the necessity of identity was provable in Ruth's system, and took this as a reductio, but he did not mention that Ruth had already proved it. The solution to this strange puzzle is that Quine wrote his review on the basis of galley proofs sent to him by Church, and "through some lapse, or possibly the loss of a second proof sheet,"[6] Quine never saw the sheet where Ruth was proving the result that Quine called for. Not only Quine, but also many modal logicians have contested Ruth's result and regard it as absurd. However, as we shall soon see, I regard it as an important insight, and the attempts by modal logicians to avoid it obfuscate their semantics.

Further, Quine argued that modal logic commits one to what he called *"Aristotelian" essentialism.* Quine means by this the doctrine that some of the attributes of a thing are essential to *it*, necessary of the thing *regardless of the way in which we refer to it,* while other attributes are accidental to it. Ruth, in three papers, Terence Parsons and others have argued against Quine. However, what they discuss under the label essentialism is something quite different from Quine's notion, which they seem to have missed. This issue concerning essentialism is one of the points where I disagree with Ruth. What she and others write about essential traits, individual essences, etc., is fine, but they are not reading Quine carefully enough and thereby miss his point. Thus, for example, one important consequence of Quine's argument is that it undercuts Carnap's *linguistic doctrine of necessity*. Much of Quine's criticism of the modalities is, explicitly or implicitly, directed against Carnap, and this is an example.[7]

Finally, in 1960, in *Word and Object*, Quine felt that he had clinched his case. He presented an argument to the effect that in quantified modal logic

---

5 Quine 1947b.

6 Quine 1958.

7 Follesdal 1986.

modal distinctions *collapse*: not only is everything that is necessary also true, but in addition, everything that is true is necessary. Given the collapse, there is no point in the modal distinctions any longer.

No modal logician seems to have taken Quine's argument seriously. There is no critical discussion of it. It seems that they thought, as they had tended to do in the case of his earlier criticism: "We *do* modal logic – where is the collapse?"

However, Quine did not claim that the collapse is *provable* in the various systems of modal logic. His claim was that the *interpretation* of modal logic will lead to a collapse of the modal distinctions. There is no way of interpreting modal logic in a clear way, ontologically and conceptually, without a collapse.

I did what every logician should do: I formalized the argument. It was very clearly spelled out in *Word and Object*, but formalizing it made explicit all assumptions of the argument. And lo and behold, all assumptions were universally shared philosophical views at the time.

Should we then give up the modalities?

No, there was an important problem with Quine's argument: It was just *too catastrophic*. Nothing was assumed about the properties of the modal operator that figured in the argument. It did not only lead to a collapse of modal distinctions, but also to a collapse of any attempt to have an operator that singles out from the class of all true sentences a *proper* subset, that is a subset that does not coincide with the whole class.

That is, Quine's argument leads to a collapse not only of the logical modalities, but also of the epistemic ones, the deontic ones, in addition to our common use of counterfactuals, probability, etc.[8] So what then remains of epistemology, ethics, and science? The argument must be wrong. But where does it go wrong?

By examining the argument I found that there was only one of its premisses that could possibly be given up: the basic assumption of all philosophy of language at that time, universally accepted by Frege, Carnap and everybody else: the view that *singular terms and general terms have basically the same semantics*: they have a sense (intension) by which they refer to their reference (extension). I called these semantics "*one-sorted semantics*." The argument turns on the freedom one has in all these one-sorted semantics to go back and forth between general terms and sentences on the one side and singular terms on the

8 The collapse of probabilities was explored by Smokler in 1977 and other works.

other, through the use of definite descriptions, class abstraction and other techniques.

Arguments based on the same idea as Quine's had been used earlier. Church[9] used an argument of this kind to argue for his Fregean semantics. Gödel[10] proposed an even simpler argument and showed that it "leads almost inevitably to the conclusion that all true sentences have the same signification (as well as all false ones)" (pp. 129–30). He noted that [I quote] "Frege actually drew this conclusion; and he meant it in an almost metaphysical sense, reminding one somewhat of the Eleatic doctrines of the 'One'." (p. 129).

Gödel presented the argument in a discussion of Russell's philosophy, and he observed that Russell's contextual elimination of definite descriptions blocks the argument. However, Gödel noticed that the argument raised serious problems for the traditional Fregean view on names and reference. He concluded that "I cannot help feeling that the problem raised by Frege's puzzling conclusion has only been evaded by Russell's theory of descriptions and that there is something behind it which is not yet completely understood." (p. 130)

A hint of what Gödel had in mind here, comes a little bit later, where he writes: "Closer examination, however, shows that this advantage of Russell's theory over Frege's subsists only as long as one interprets definitions as mere typographic abbreviations, not as introducing names for objects described by the definitions, a feature which is common to Frege and Russell." (p. 131). So what Gödel was missing, was a semantics for names that gets around his argument.

In my dissertation, I proposed such a way of handling names: I argued that names do not philander from object to object the way definite descriptions do, but that they stick to the same object in all possible worlds. Names in ordinary language do not behave like disguised descriptions. I wrote in my dissertation about the view that names stick to the same object in all possible worlds: "this solution leads us *to regard a word as a proper name of an object only if it refers to this one and same object in all possible worlds*." And I continue: "This does not seem unnatural. Neither does it seem preposterous to assume as we just did, that if a name-like word does not stick to one and the same object in all possible worlds, the word contains some descriptive element."

As far as I know, Gödel was the only one who had regarded the argument with suspicion before it was discussed critically in this dissertation. My diagno-

---

9 Church 1943, esp. pp. 299–300.

10 Kurt Gödel (1944), esp. pp. 128–131 (= pp. 122–124 of the reprint in Solomon Feferman et al., eds.).

sis is, as mentioned, that in the argument one slides too easily between general terms and singular terms. The proposed two-sorted semantics prevents such a slide and does justice to the behavior of names in natural languages.

In the years that have followed, a large number of articles and books have been devoted to the argument, often repeating earlier work. A notable contribution was made by Barwise and Perry in 1981.[11] The very apt name "the slingshot" for this argument and similar ones is due to them.

Where, then, do we stand? I think we have to accept a package having the following four ingredients:

1. **Necessity of identity**. In my view, Ruth's result of 1946, although rejected as absurd by many critics, is an important feature of this package. Quine's analysis of the situation gives further support to this view, although Quine himself, sticking to the traditional Fregean semantics, regarded it as absurd.
2. **Aristotelian essentialism**. Quine's analysis shows that quantification into modal contexts requires Aristotelian essentialism in order to be intelligible. For Quine, this was an argument against Carnap and his linguistic doctrine of necessity, and rightly so, that doctrine is incoherent. However, if we want to keep some of the basic notions of science, like counterfactuals and probability, we have to accept that "necessity resides in things, not in language."
3. **Give up the slingshot**. The slingshot is, as we have noted, fatal not just to the logical modalities, but to epistemology, ethics and science as well. So we should give it up. But the way of giving it up that I favor, is to give up the traditional one-sorted semantics in favor of a two-sorted semantics where singular terms have a semantics quite different from general terms and sentences.

This leads us to the fourth ingredient in our package:

4. **Names have to refer to the same object in all possible worlds.**
However, this fourth ingredient can be achieved in different ways. At the meeting at Harvard in 1962, to which Ruth often refers, she argued that names are to be considered as *tags*, not involving any descriptive content. However, she concludes her paper by proposing a "substitutional" interpretation of the singular terms (where, roughly, an existentially quantified

11 Barwise and Perry, pp. 387–404.

> sentence is true if and only if there is a name in the language which when substituted for the quantified variable makes the sentence true). Further, early and late she subscribed to Arthur F. Smullyan's use of Russell's theory of descriptions as a way out of Quine's quandaries concerning the modalities. Smullyan seems to hold that while names and descriptions do behave differently in modal contexts, this difference in behavior is explicable by the thesis that names are descriptions which have widest scope. This is definitely different from the view I put forth in my dissertation. I do not regard names as disguised descriptions, but as a special category of expressions which signal that we want to keep on referring to the same object in extensional contexts as well as in modal contexts. And I regard a substitutional interpretation of quantification as a wrong way out. Reference, both of variables of quantification and of names, is to objects. And proper names reflect the important role that objects play in our lives and in our communication. Today's symposium may be an occasion for clarifying where we agree and where we disagree.

As for Kripke's view, in the discussion following Ruth's lecture in 1962 he seemed to agree with her. He stated:

> the assumption of a distinction between tags and empirical descriptions, such that the truth-values of identity statements between tags (but not between descriptions) are ascertainable merely by recourse to a dictionary, amounts to essentialism itself. The tags are the "essential" denoting phrases for individuals, but empirical descriptions are not, and thus we look to statements containing "tags", not descriptions, to ascertain the essential properties of individuals. Thus the distinction between "names" and "descriptions" is equivalent to essentialism.[12]

As I noted earlier, my discussion of essentialism in my dissertation does not have to do with names, essentialism comes in already when we quantify into modal contexts: "To make sense of Aristotelian essentialism and to make sense of open sentences with an '□' prefixed are one and the same problem, and a solution to the one is a solution to the other," I wrote.[13] This is also what Quine notes in his answer to Kripke:

12 Page 115 of the transcript in Wartofsky.

13 Føllesdal 1961, 2004, Section 19, page 92.

> My answer is that this kind of consideration is not relevant to the problem of essentialism because one doesn't ever need descriptions or proper names.[14]

It may be that we here have the beginning of the different notion of essentialism that I mentioned above that became prominent in Kripke's work and was rejected by Ruth. A further problem of Kripke's observation is that if names are supposed to refer in virtue of essential properties, then we are back to regarding names as a variety of descriptions again. However, it is not clear whether this is what Kripke had in mind here. And there is no indication that Ruth shared his view. Nine years later, in 1971, Kripke had developed a different view and he now argued for what he aptly calls "rigid designators" by surveying the behavior of names in ordinary language.

My own view in 1961 and also now is different from Kripke's. I argued that what I called "genuine singular terms," including proper names, are referring to the same object in all possible worlds, but I did not regard this as due to their attaching to some essence or due to their being mere tags. What, following Quine, I call "Aristotelian essentialism" is the view that causality and other modalities depend on features of things and not on the language we use to talk about these things. These features are not "individual essences," several things may have the same features. And names and other referring expressions do not refer to their objects in virtue of some unique essential features specific to each object. A main point of my thesis was that we have to accept the whole package that I just outlined, necessity of identity, Aristotelian essentialism, rejection of the slingshot, and a two-sorted semantics, where the proper, or genuine, singular terms keep their reference in all possible worlds.

Ruth describes in her book *Modalities* the discussion at Harvard in 1962 as if she were in a lion's den, where she appreciated Saul Kripke's support. She clearly believed that being a student of Quine I sided with Quine in his rejection of the modalities. In "A Backward Look at Quine's Animadversions on Modalities" (1990) she writes:

> Føllesdal in his 1986 paper for the Quine volume of the Library of Living Philosophers now endorses the special role of proper names and a theory of direct reference. There is no longer a problem of "making sense" of expressions with modal operators attached to open sentences or modal operators in the scope of quantifiers. ... We have come a long way.[15]

---

14 Page 115 of the transcript in Wartofsky.

15 Barcan Marcus 1993, p. 232

The "now" here is clearly misleading. I had this view already in 1961, a year before Ruth gave her talk at Harvard. I still do not have a satisfactory view on how names relate to their objects. However, the points Ruth mentions were there already in 1961.

Indeed, Ruth and I agree on many of the basic issues concerning the interpretation of the modalities. She was the first to take the very important step of developing a quantified modal logic, and she saw very early, in 1947, that identity statements are necessary. She might have been relieved in 1962, when she visited Harvard, if she had known that she had one more ally in that group, who had the year before argued for and accepted the whole package that I have outlined, and who had absolutely no "problem of 'making sense' of expressions with modal operators attached to open sentences or modal operators in the scope of quantifiers."

I have always, ever since I was a student, appreciated Ruth's path-breaking work. It is reflected in the very many references I make to her in my thesis. I regard it as a basic principle in all research and scholarship that one should refer to all who have presented the same or similar ideas previously. In the case of Ruth there was a lot of important and innovative work to refer to.

She has written much since then, and I continue to appreciate her work. I also appreciate her warm friendship over all these years, and I now congratulate her with the Lauener Prize. It is very much deserved.

References

Barcan, Ruth C. (1946a) "A Functional Calculus of First Order Based on Strict Implication." *Journal of Symbolic Logic* 11, pp. 1–16.

Barcan, Ruth C. (1946b) "The Deduction Theorem in a Functional Calculus of First Order Based on Strict Implication." *Journal of Symbolic Logic* 11, pp. 115–118.

Barcan, Ruth C. (1947) "The Identity of Individuals in a Strict Functional Calculus of Second Order." *Journal of Symbolic Logic* 12, pp. 12–15.

Barcan, Ruth C. (1948) Review of Smullyan's "Modality and Description." *Journal of Symbolic Logic* 13, pp. 149–150. Reprinted as "Smullyan on Modality and Description" in Barcan Marcus 1993, pp. 36–38.

Barcan Marcus, Ruth (1961) "Modalities and Intensional Languages," *Synthese* 13, 303–322. Reprinted in Barcan Marcus 1993, pp. 5–23.

Barcan Marcus, Ruth (1967) "Essentialism in Modal Logic," *Nous* 1, 90–96. Reprinted in Barcan Marcus 1993, pp. 45–51.

Barcan Marcus, Ruth (1990) "A Backward Look at Quine's Animadversions on Modalities." In Robert B. Barrett and Roger F. Gibson, eds., *Perspectives on Quine*, Oxford and Cambridge, Mass.: Blackwell, pp. 230–243. Reprinted in Barcan Marcus 1993, pp. 216–232.

Barcan Marcus, Ruth (1993) *Modalities: Philosophical Essays*. Oxford: Oxford University Press.
Barwise, Jon, and John Perry (1981) "Semantic Innocence and Uncompromising Situations." *Midwest Studies in Philosophy* 6, pp. 387–404.
Carnap, Rudolf (1946) "Modalities and Quantification." *Journal of Symbolic Logic* 11, pp. 33–64.
Carnap, Rudolf (1947, 1956) *Meaning and Necessity*. Chicago: University of Chicago Press. 2nd ed., with supplements, 1956.
Church, Alonzo (1943) Review of Carnap's *Introduction to Semantics*. *Philosophical Review* 52, pp. 298–304.
Føllesdal, Dagfinn (1961, 2004) *Referential opacity and modal logic*. (Thesis for the Ph.D. degree, Harvard). London: Routledge, 2004.
Føllesdal, Dagfinn (1986) "Essentialism and reference", Lewis E. Hahn and Paul Arthur Schilpp, eds., *The Philosophy of W.V. Quine* (The Library of Living Philosophers), La Salle, Ill., pp. 97–113.
Føllesdal, Dagfinn (1994) "Stig Kanger in Memoriam." In *Logic, Methodology and Philosophy of Science* IX, edited by Dag Prawitz, Brian Skyrms and Dag Westerståhl. Amsterdam: Elsevier Science, pp. 885–888.
Gödel, Kurt (1944) "Russell's mathematical logic," In P. A. Schilpp (ed.), *The Philosophy of Bertrand Russell* (Library of Living Philosophers), Evanston, Ill.: Northwestern University Press, pp. 123–153. Reprinted in several collections, including S. Feferman, John W. Dawson, Stephen C. Kleene, G. Moore, R. Solovay and Jean van Heijenoort (eds), *Kurt Gödel, Collected Works*,Volume II: Publications 1938–1974. Oxford: Oxford University Press, 1990, pp. 119–141.
Hintikka, K. Jaakko J. (1957a) "Modality as Referential Multiplicity." *Ajatus* 20, pp. 49–64.
Hintikka, K. Jaakko J. (1957b) "Quantifiers in Deontic Logic." *Societas Scientiarum Fennica, Commentationes Humanarum Literarum* 23, no. 4. Helsinki.
Kanger, Stig (1957) *Provability in Logic* (Stockholm Studies in Philosophy 1). Stockholm: Almqvist & Wiksell.
Kripke, Saul A. (1971) "Identity and Necessity." In Milton K. Munitz (ed.), *Identity and Individuation*, New York: New York University Press, pp. 135–164.
Kripke, Saul A. (1972) "Naming and Necessity," *Synthese* 40, 253–354. Expanded version: *Naming and Necessity*, Cambridge, Mass.: Harvard University Press, 1980.
Parsons, Terence (1967) "Grades of Essentialism in Modal Logic," *Nous* 1, 181–191.
Parsons, Terence (1969) "Essentialism and Quantified Modal Logic," *Philosophical Review* 78, 32–52.
Quine, Willard Van Orman (1946) Review of Barcan's "A Functional Calculus of First Order Based on Strict Implication." *Journal of Symbolic Logic* 11, pp. 96–97.
Quine, Willard Van Orman (1947a) Review of Barcan's "The Deduction Theorem in a Functional Calculus of First Order Based on Strict Implication." *Journal of Symbolic Logic* 12, p. 95.
Quine, Willard Van Orman (1947b) Review of Barcan's "The Identity of Individuals in a Strict Functional Calculus of Second Order." *Journal of Symbolic Logic* 12, pp. 95–96.
Quine, Willard Van Orman (1958) Correction to the review of Barcan's "The Identity of Individuals in a Strict Functional Calculus of Second Order." (Quine 1947b above). *Journal of Symbolic Logic* 23, p. 342.
Quine, Willard Van Orman (1960) *Word and Object*. Cambridge, Mass.; MIT Press.
Smokler, Howard (1977) "Three grades of probabilistic involvement." *Philosophical Studies* 32, pp. 129–142.

Wartofsky, Marx, ed. (1963) *Boston Studies in the Philosophy of Science*, pp. 77–96. Dordrecht, Holland: Reidel.

Timothy Williamson

Barcan Formulas in Second-Order Modal Logic

Abstract

Although the variable-domain Kripke semantics for first-order modal logic invalidates the first-order Barcan schema and its converse, its natural extension to second-order modal logic *validates* the second-order analogues of those schemas, where the second-order quantifiers are interpreted as ranging over intensions, even if the intensions are restricted by a 'serious actualist' constraint. The same semantics validates a strong comprehension principle for the second-order quantifiers, which has instances that are hard to reconcile with the contingency of being implied by the rejection of the first-order Barcan schema and its converse, on any metaphysically plausible view. A defender of the contingency of being may reject the Kripke semantics but still needs some alternative comprehension principle for the second-order quantifiers that avoids the problematic instances. The paper considers various salient candidates for such a weakened comprehension principle, but shows that in different ways they are too weak to generate a satisfactory second-order modal logic. The provisional conclusion of the paper is that since second-order modal logic, on a legitimate intensional interpretation, requires a comprehension principle that is defensible only if the contingency of being is rejected, we should indeed reject the contingency of being, and accept both the first-order and second-order versions of the Barcan schema and its converse. A corresponding metaphysical conception is briefly sketched.

University of Oxford

Earlier versions of this material were presented at a workshop on philosophical logic at Oxford University, colloquia at Texas A&M University, Rice University and Peking University, and of course at the 2008 International Lauener Symposium on Analytical Philosophy in honour of Ruth Barcan Marcus in Berne. I thank participants in all these events for helpful questions, and Øystein Linnebo, Philip Percival and Gabriel Uzquiano for detailed written comments. An AHRC Research Leave Award provided funding for research on the topic of this paper. A later and more developed version of the argument appears in Williamson (2013), especially chapter 6, parts of which overlap the present material and were derived from it.

Second-order logic and modal logic are both, separately, major topics of philosophical discussion. Although both have been criticized by Quine and others, increasingly many philosophers find their strictures uncompelling, and regard both branches of logic as valuable resources for the articulation and investigation of significant issues in logical metaphysics and elsewhere. One might therefore expect some combination of the two sorts of logic to constitute a natural and more comprehensive background logic for metaphysics. So it is somewhat surprising to find philosophical discussion of second-order modal logic scarce, despite the pioneering contribution of Barcan (1947).

Two contrary explanations initially suggest themselves. One is that the topic of second-order modal logic is too hard: multiplying together the complexities of second-order logic and of modal logic produces an intractable level of technical complication. The other explanation is that the topic is too easy: its complexities are just those of second-order logic and of modal logic separately, combining which provokes no special further problems of philosophical interest. These putative explanations are less opposed than they first appear, since some complexity is boring and routine. Nevertheless, separately and even together they are not fully satisfying. For the technical complexities of second-order modal logic are no worse than those of many other branches of logic to which philosophers appeal: the results in this paper are proved in a few lines. Nor are the complexities philosophically unrewarding. As we shall see, the interaction of second-order quantifiers with modal operators raises deep issues in logical metaphysics that cannot be factorized into the issues raised by the former and the issues raised by the latter.

Such fruitful interaction is already present in the case of first-order modal logic. The Barcan formula, introduced in Barcan (1946), raises fundamental issues about the contingency or otherwise of existence, issues that arise neither in first-order non-modal logic nor in unquantified modal logic. For second-order modal logic there are both first-order and second-order Barcan formulas. Perhaps surprisingly, the issues about the status of the second-order Barcan formula are *not* simply higher-order analogues of the issues about the status of the first-order Barcan formula. Nevertheless, reflection on the status of the second-order Barcan formula and related principles casts new light on the controversy about the status of the first-order Barcan formula. We begin by sketching some of the issues around the first-order Barcan formula, before moving to the second-order case.

1. A standard language L1 for first-order modal logic has countably many individual variables **x**, **y**, **z**, ..., an appropriate array of atomic predicates (non-

logical predicate constants and the logical constant =), the usual truth-functors (¬, **&**, ∨, →, ↔), modal operators (◊, □) and first-order quantifiers (∃, ∀). Of those operators, ¬, **&**, ◊ and ∃ are treated as primitive. In what follows, we have in mind readings of the modal operators on which they express metaphysical possibility and necessity respectively.

The Barcan formula is really a schema with infinitely many instances. Contraposed in existential form it is:

BF $\quad \Diamond\exists \mathbf{x}\, \mathbf{A} \rightarrow \exists \mathbf{x}\, \Diamond \mathbf{A}$

Here **x** is any variable and **A** any formula, typically containing free occurrences of **x** (and possibly of other variables). We can informally read BF as saying that if there could have been an object that met a given condition, then there is an object that could have met the condition. We also consider the converse of the Barcan formula:

CBF $\quad \exists \mathbf{x}\, \Diamond \mathbf{A} \rightarrow \Diamond\exists \mathbf{x}\, \mathbf{A}$

We can informally read CBF as saying that if there is an object that could have met the condition, then there could have been an object that met the condition.

Any philosophical assessment of BF and CBF must start by acknowledging that there *seem* to be compelling counterexamples to both of them. The counterexamples flow naturally from a standard conception of existence as thoroughly contingent, at least in the case of ordinary spatiotemporal objects.

For BF, read **A** as '**x** is a child of Ludwig Wittgenstein' (in the biological sense of 'child'). Then the antecedent of BF says that there could have been an object that was a child of Ludwig Wittgenstein. That is true, for although Wittgenstein had no child, he could have had one. On this reading, the consequent of BF says that there is an object that could have been a child of Ludwig Wittgenstein. That seems false, given plausible-looking metaphysical assumptions. For what is the supposed object? It is not the child of other parents, for by the essentiality of origin no child could have had parents other than its actual ones. Nor is it a collection of atoms, for although such a collection could have *constituted* a child, it could not have been *identical* with a child. There seems to be no good candidate to be the supposed object. Thus BF seems false on this reading.

For CBF, read **A** as '**x** does not exist', in the sense of 'exist' as 'be some object or other'. Then the antecedent of CBF says that there is an object that could have not existed. That seems true: it seems that each one of us is such an object. For example, my parents might never have met, and if they had not I would never have existed; I would not have been any object at all. On this reading, the consequent of CBF says that there could have been an object that did not exist. That is false; there could not have been an object that was no object at all. Thus CBF seems false on this reading.

Kripke (1963) provided a formal semantics (model theory) for first-order modal logic that invalidates BF and CBF and thereby appears to vindicate the informal counterexamples to them. Here is a very slightly revised version of his account. A model is a quintuple $\langle W, w_0, D, \text{dom}, \text{int}\rangle$ where W and D are non-empty sets, $w_0 \in W$, dom is a function mapping each $w \in W$ to $\text{dom}(w) \subseteq D$, and int is a function mapping each non-logical n-place atomic predicate **F** to a function int(**F**) mapping each $w \in W$ to $\text{int}(\mathbf{F})(w) \subseteq \text{dom}(w)^n$. Informally, we can envisage W as the set of possible worlds, of which w_0 is the actual world, dom(w) as the set of objects that exist in the world $w \in W$, int(**F**) as the intension of **F** and int(**F**)(w) as the extension of **F** with respect to w. However, these informal glosses play no essential role in the formal model theory itself.

Several features of the models are worth noting.

First, it is a variable domains model theory: different domains can be associated with different worlds. This is crucial to the formal counter-models to BF and CBF, and reflects the conception of existence as a thoroughly contingent matter.

Second, the extension of each atomic predicate in a world comprises only things that exist in that world; in this sense the model theory respects what is sometimes called 'serious actualism'. An object cannot even be self-identical with respect to a world in which it does not exist, because there is nothing there to be self-identical. This feature of the models is imposed here in order not to undermine the conception of existence as thoroughly contingent, since to put a non-existent object into the extension of a predicate is hardly to take its non-existence seriously. In his 1963 paper, Kripke describes the serious actualist constraint as 'natural', but refrains from imposing it in order to maintain the rule of uniform substitution. For present purposes, the philosophical motivation for imposing the constraint matters more. In any case, if we count a complex open formula such as ¬**Fx** as no substitution instance of **Fx**, on the grounds that sentence operators such as ¬ are not predicate operators, then uniform substitution remains valid, albeit on a somewhat emasculated reading, as discussed by Stalnaker (1977).

Third, although the set D from which members of the domains of worlds are drawn is nonempty — to ensure that values can be assigned to the individual variables — it is not required that individuals exist in any world. Models with empty worlds are allowed; even models in which all worlds are empty are allowed.

Fourth, the models do not include an accessibility relation R that would enable a restriction of the semantic clauses for the modal operators to accessible worlds. Consequently, at the propositional level they validate the strong modal logic S5, in which all necessities are necessarily necessary and all possibilities are necessarily possible. The accessibility relation is omitted only for simplicity; it could easily be added if desired.

We now define what it is for a formula of L1 to be true in a model. As usual, we first define the truth of a formula relative to an assignment at a world in a model. We assume a model $<W, w_0, D, dom, int>$ given and leave reference to it tacit. An assignment is a function from all variables to members of D. '$w, a \models \mathbf{A}$' means that the formula **A** is true at $w \in W$ on assignment a. We define this relation recursively, letting **F** be an n-place atomic predicate, $\mathbf{v}, \mathbf{v}_1, \ldots, \mathbf{v}_n$ (first-order) variables and $a[\mathbf{v}/o]$ the assignment like a except that it assigns o to **v**:

$w, a \models \mathbf{Fv_1 \ldots v_n}$ iff $<a(\mathbf{v}_1), \ldots, a(\mathbf{v}_n)> \in int(\mathbf{F})(w)$

$w, a \models \mathbf{v_1{=}v_2}$ iff $<a(\mathbf{v}_1), a(\mathbf{v}_2)> \in \{<o, o>: o \in dom(w)\}$

$w, a \models \neg\mathbf{A}$ iff not $w, a \models \mathbf{A}$

$w, a \models \mathbf{A \,\&\, B}$ iff $w, a \models \mathbf{A}$ and $w, a \models \mathbf{B}$

$w, a \models \exists\mathbf{v}\,\mathbf{A}$ iff for some $o \in dom(w)$: $w, a[\mathbf{v}/o] \models \mathbf{A}$

$w, a \models \Diamond\mathbf{A}$ iff for some $w^* \in W$: $w^*, a \models \mathbf{A}$

A is true at w if and only if for all assignments a: $w, a \models \mathbf{A}$. **A** is true in the model if and only if it is true at w_0. **A** is valid if and only if it is true in all models.

A formula **A** can be true at the actual world of a model without being true at every world in the model; in that case, **A** is true in the model but $\Box\mathbf{A}$ is not. However, if **A** is untrue at w in the model $<W, w_0, D, dom, int>$ then **A** is also untrue at w in the distinct model $<W, w, D, dom, int>$, and therefore untrue in $<W, w, D, dom, int>$. Contrapositively, if **A** is valid, then $\Box\mathbf{A}$ is also valid.[1]

---

1 The point depends on the fact that the selection of one member of W as the actual world plays no role in the definition of truth at a world in the model. If the language included a rigidifying 'actually' operator, the selected world would play a distinctive in its semantic clause, and

Indeed, if **A** is valid, so is any closure of **A**, that is, any result of prefixing **A** by universal quantifiers and necessity operators in any order. It will be convenient in what follows to treat any closure of an instance of a schema (such as BF or CBF) as itself an instance of that schema. For instance, $\Box(\Diamond\exists x\, A \rightarrow \exists x\, \Diamond A)$ will count as an instance of BF. A schema is valid if and only if all its instances are valid; it is valid in a model if and only if all its instances are true in that model.

It is a purely mathematical exercise to show that BF and CBF are invalid on the Kripke semantics, by providing a model in which instances of them are not true. The same counter-model will do for both. Consider $<W, w_0, D, \mathrm{dom}, \mathrm{int}>$, where $W = \{0, 1\}$, $w_0 = 0$, $D = \{2, 3\}$, $\mathrm{dom}(0) = \{2\}$, $\mathrm{dom}(1) = \{3\}$. For BF, let $a(\mathbf{y}) = 3$ and observe that $0, a \models \Diamond\exists x\, x{=}y$ but not $0, a \models \exists x\, \Diamond x{=}y$; thus not $0, a \models \Diamond\exists x\, x{=}y \rightarrow \exists x\, \Diamond x{=}y$, so that instance of BF is not true in the model. For CBF, observe that $\exists x\, \Diamond\neg x{=}x$ but not $\Diamond\exists x\, \neg x{=}x$ is true at 0; thus $\exists x\, \Diamond\neg x{=}x \rightarrow \Diamond\exists x\, \neg x{=}x$ is not true in the model. Such counter-models to BF and CBF look like formal analogues of informal counter-examples to them such as those presented above.

Kripke established general correspondence results between the structure of models and the validity of BF and CBF. For models of the present simple sort, lacking the accessibility relation R, his results boil down to this: both BF and CBF are valid in a model where all worlds have the same domain; both BF and CBF are invalid in a model where not all worlds have the same domain.[2] We seem to have compelling reason not to impose the restrictions on models required for the validity of BF and CBF: since there could have existed an object that does not actually exist (such as a child of Wittgenstein), something may exist in some world without existing in the actual world; since there exists an object that could have failed to exist, something may exist in the actual world without existing in every world.

On further reflection, the case against BF and CBF looks much less solid. Consider first the role of the Kripke semantics. Obviously, the mere mathematical fact that BF and CBF are invalid over some class of formal models by itself shows nothing about whether they have false instances on their intended interpretation, on which the symbol $\Diamond$ expresses metaphysical possibility. The question is how the formal models correspond to the intended interpretation.

---

the validity of **A** would no longer imply that of $\Box$**A**. That does not affect the arguments in this paper.

2 In this simple setting, BF is valid in a given model if and only if CBF is. In more complex settings, where an accessibility relation is introduced or the necessitations of instances of BF and CBF are not themselves counted as instances of them, either schema may be valid in a model when the other is not. Those complications too do not affect the arguments of this paper.

The problem is most immediate for a Kripke counter-model to an unnecessitated instance of BF. Such a model must have some $w \in W$ and some $o \in \text{dom}(w)$ such that $o \notin \text{dom}(w_0)$. But, on the intended interpretation, w_0 is the actual world and $\text{dom}(w_0)$ contains everything that exists in the actual world, in other words, whatever there actually is — and whatever there is, there actually is. For on the metaphysically relevant readings of BF and CBF, their quantifiers are not restricted by some property of existence that excludes some of what there is.[3] But if there is such a counter-model, then there is such an element o of its domain D, so there is such an object o, so it should be that $o \in \text{dom}(w_0)$ after all, contrary to hypothesis. Therefore, no Kripke counter-model to unnecessitated BF is an intended model — even though the proposed informal counterexamples concern such unnecessitated instances. Someone might still claim that a Kripke counter-model to unnecessitated BF somehow formally represents a genuine counter-example to an unnecessitated instance of BF on its intended interpretation: some but not all of the objects there are would formally represent all of the objects there are. But the mere existence of the formal representation itself would not constitute any positive reason to think that there really was a counter-example to unnecessitated BF on its intended interpretation. The existence of a Kripke counter-model to BF is an elementary non-modal mathematical fact; it is no evidence for the modal claim that there could have been some objects other than all the actual objects. The Kripke semantics provides no objection to unnecessitated BF independent of the apparent informal counter-instances. It merely provides an elegant and tractable formal representation of the structure of those apparent counter-instances, and serves as a useful algebraic instrument for establishing mathematical results about quantified modal logic, such as the independence of BF from various other principles.

The same moral applies to CBF and necessitated BF, even though the problem for Kripke counter-models to them is less blatant than for unnecessitated BF. The attempt to construe the counter-model as intended does not run into the same immediate contradiction, since CBF and necessitated BF have counter-models in which $\text{dom}(w_0)$ is the whole of D; then we only need $\text{dom}(w)$ to be a

---

3 We will therefore not be concerned with the denial of BF or CBF by modal realists such as David Lewis, since that depends on a restriction of the quantifier to what is in the given world in a literal sense of 'in'. For Lewis, our unrestricted quantifiers range over everything in every world, as do those of speakers in other worlds. He provides no conception of contingent existence in the most interesting, radical sense.

proper subset of D for some counterfactual world *w*.[4] Nevertheless, the mere existence of Kripke counter-models to CBF and necessitated BF is just an elementary non-modal mathematical fact. It is no evidence that there really is a counter-instance to CBF or necessitated BF on its intended interpretation, that some objects could have failed to exist. Again, the Kripke semantics provides no objection to CBF or necessitated BF independent of the apparent informal counter-instances.

Thus the whole weight of the objection to BF and CBF falls on the informal putative counter-instances. But they too establish less than they seemed to at first sight. For BF, **A** was read as '**x** is a child of Ludwig Wittgenstein'. On this reading, the antecedent of BF is obviously true, but its consequent is not obviously false. There is indeed nothing concrete that could have been a child of Wittgenstein, but that does not eliminate the alternative that there is something non-concrete that could have been a child of Wittgenstein (in which case it would have been concrete). Such a contingently non-concrete object is a possible child of Wittgenstein not in the sense of being a child of Wittgenstein that contingently fails to exist (necessarily, every child of Wittgenstein has concrete existence) but in the sense of being something that could have been a child of Wittgenstein. For CBF, **A** was read as '**x** does not exist', in the sense of 'exist' as 'be some object or other'. On this reading, the consequent of CBF is obviously false, but its antecedent is not obviously true. There could indeed have failed to be any such concrete object as you, but that does not prove that there could have failed to be any such concrete or non-concrete objects as you; perhaps you could have been a contingently non-concrete object. Thus an ontology that allows for contingently non-concrete objects is compatible with the conjunction of BF and CBF. It can explain away the apparent counter-instances to them as based on a neglect of that category.[5]

However, it is one thing to specify a consistent ontology on which BF and CBF hold, quite another to provide positive reason to accept that ontology. Why should we think that there can be contingently non-concrete objects? Elsewhere, I have given some tentative philosophical arguments for such an

4 The problem remains that by Russell's paradox there are too many actual objects (such as sets) to constitute a set, whereas dom(w_0) must be a set; but this is a quite general difficulty for the notion of an intended model in a set-theoretic framework, and has no special connection with modal issues. See Williamson (2000a) for discussion.

5 Such a defence of BF and CBF is provided in Williamson (1990, 1998, 2000b), Linsky and Zalta (1994, 1996) and Parsons (1995). Marcus (1985/1986) suggests a different sort of defence of BF.

ontology.[6] Furthermore, first-order modal logic with BF and CBF is technically simpler and more streamlined than first-order modal logic without them, so considerations of systematicity tell in favour of BF and CBF.[7] The present paper comes at the issue from a different angle, by assessing the status of BF and CBF in second-order modal logic. Basic forms of second-order reasoning turn out to be hamstrung without a strong comprehension principle that is hard to reconcile with the rejection of BF and CBF in any metaphysically plausible way.

2. Suppose that Alice does not smoke, although she could have smoked. Then there is something that Alice does not do, although she could have done it. The simplest formal counterpart of that valid argument involves quantification into predicate position, with a premise of the form $\neg \mathbf{Sa} \;\&\; \Diamond \mathbf{Sa}$ and a conclusion of the form $\exists \mathbf{X}\,(\neg \mathbf{Xa} \;\&\; \Diamond \mathbf{Xa})$, where the second-order variable **X** occupies the position of the monadic predicate **S**. We should not think of second-order variables as restricted to 'genuine properties' that are fundamental in physics or imply significant similarity between their exemplars. The property of smoking is not fundamental in physics, and if this is green, that is red and the other is blue then in the relevant sense it follows that there is something that this and that are but the other is not, namely red or green, even though it does not imply significant similarity between its exemplars. Some of the most important uses of second-order logic are mathematical, where second-order quantification is needed to capture the intended interpretation of the principle of mathematical induction, the definition of the ancestral of a relation and the separation principle about the existence of sets: in many mathematical applications, the predicates fed into those principles are not guaranteed to express 'genuine properties'. Thus we should read the second-order quantification as plenitudinous, not sparse. Any predicate will do to fix a value for a second-order variable. In the standard model theory for second-order logic, this idea is captured by having the second-order quantifiers range over all subsets of the domain over whose members the first-order quantifiers range.[8]

We can extend the standard model theory of second-order logic to the modal case, by using Kripke models. As already seen, the opponent of BF and CBF should assign an instrumental role to Kripke models, rather than taking any of them to capture the intended interpretation of the language. Still, they

6 See Williamson (1998, 2000a, 2002).

7 See Cresswell (1991), Linsky and Zalta (1994), and Williamson (1998).

8 For the model theory of second-order logic see Shapiro (1991).

give clues as to which principles of second-order modal logic may be expected to hold. We shall later see how to do without them.

We expand the first-order language L1 to a second-order language L2 by adding countably many variables **X**, **Y**, **Z**, ... that take the syntactic position of 1-place atomic predicates.[9] With the usual harmless ambiguity, we use the same symbol (∃) for first-order and second-order quantifiers.

In a standard model for a second-order non-modal language, the domain of the first-order quantifiers fixes the domain of the second-order quantifiers, so the latter requires no independent specification. Thus a model for a first-order non-modal language serves equally well as a standard model for the corresponding second-order non-modal language.[10] The same holds in the modal case.[11] A Kripke model is a quintuple <W, w_0, D, dom, int>, just as before.

An n-place intension is any function f mapping each $w \in W$ to $f(w) \subseteq \text{dom}(w)^n$. Since the original Kripke semantics simply associated each n-place atomic predicate **F** with an intension int(**F**), and the second-order variables occupy the position of 1-place predicates, in the new semantics we simply require an assignment a to map each second-order variable **V** to a 1-place intension a(**V**), in addition to assigning values from D to the first-order variables as before. The use of intensions would not have been plausible if we had been using the plural interpretation of the second-order quantifiers (Boolos 1984), for the latter requires extensionality: if every one of these things is one of those things and *vice versa* then these things *just are* those things, and nothing could have been one of these things without being one of those things or *vice versa*, whereas two intensions can coincide in (nonempty) extension at one world without coinciding in extension at another.[12] But that is exactly right for properties: two properties can coincide in extension at one world without coinciding in extension at another.

---

9 We could add n-place predicate variables for each n, but for present purposes that complication is unnecessary.

10 For reasons explained in Williamson (2003), the talk of 'values' of second-order variables (whether they are sets or objects of some other kind) constitutes an undesirable and unnecessary reduction of the second-order to the first-order, by doing the semantics of a second-order language in a first-order meta-language. For present purposes such talk is harmless; we will not take the trouble to eliminate it.

11 A recent introduction to higher-order modal logic is Muskens (2007), although its focus is mainly on non-standard models.

12 For second-order modal logic and the problems it raises for BF and CBF under the plural interpretation see Williamson (2013).

All the original semantic clauses remain unchanged. We add the obvious semantic clause for second-order variables in atomic formulas:

$w, a \models \mathbf{Vv}$ iff $a(\mathbf{v}) \in a(\mathbf{V})(w)$

We also add the obvious semantic clause for the second-order quantifier:

$w, a \models \exists\mathbf{V}\ \mathbf{A}$ iff for some 1-place intension I: $w, a[\mathbf{V}/\mathrm{I}] \models \mathbf{A}$

Strikingly, this standard semantics validates the second-order versions of BF and CBF, even though it invalidates their first-order versions as before (since the domains of the first-order quantifiers are still allowed to vary between worlds):

BF2 $\Diamond\exists\mathbf{X}\ \mathbf{A} \rightarrow \exists\mathbf{X}\ \Diamond\mathbf{A}$

CBF2 $\exists\mathbf{X}\ \Diamond\mathbf{A} \rightarrow \Diamond\exists\mathbf{X}\ \mathbf{A}$

Since the intensions over which the second-order quantifiers range are restricted to those that for each world deliver a subset of its first-order domain as the extension, they are sensitive to the variability of the first-order domains. For instance, the intension corresponding to self-identity delivers at each world the first-order domain of that world as the extension, so those extensions vary exactly as much as the first-order domains. However, that cross-world variation in extension *within* an intension induces no cross-world variation in the domain of the second-order quantifiers (the set of 1-place intensions). As is visible in the semantic clause for the second-order quantifier, the restriction on the set of intensions is independent of the world of evaluation: the parameter '*w*' does not occur in the phrase 'for some 1-place intension I', whereas it occurs essentially in the corresponding phrase 'for some $o \in \mathrm{dom}(w)$' in the semantic clause for the first-order quantifier. In a quite natural way, without *ad hoc* stipulation, the second-order quantifier has a fixed domain even though the first-order domain does not.[13] Consequently, BF2 and CBF2 are validated, even though BF and CBF are not. Metaphorically, it is not just that the value of a second-order variable can be used to describe every possible world from an external perspective, as the opponent of BF and CBF may maintain with respect to the value of a first-order variable: the value of the second-order variable is present *in* every

13 See also Parsons (1983), p. 336.

possible world, because it is there available to be quantified over, unlike a contingently non-existent object.

The semantics also validates this strong comprehension principle:

CP⁺ $\exists\mathbf{X}\, \Box\forall\mathbf{x}\, (\mathbf{Xx} \leftrightarrow \mathbf{A})$

Here **A** is a formula in which any first-order variable and any second-order variable other than **X** can occur free.[14]

A tension begins to emerge between the strong comprehension principle and the failure of first-order BF and CBF. Since (1) is a closure of CP⁺, it too is valid:

(1) $\Box\forall\mathbf{y}\, \Box\exists\mathbf{X}\, \Box\forall\mathbf{x}\, (\mathbf{Xx} \leftrightarrow \mathbf{A})$

Let **A** be $\neg\mathbf{x}=\mathbf{y}$. Thus (2) counts as a valid instance of (1) and CP⁺:

(2) $\Box\forall\mathbf{y}\, \Box\exists\mathbf{X}\, \Box\forall\mathbf{x}\, (\mathbf{Xx} \leftrightarrow \neg\mathbf{x}=\mathbf{y})$

Loosely paraphrased, (2) says that negative haecceities have necessary existence: thus there would have been the property of not being me, even if there had not been me. Necessarily, if there is me, then everything but me has that property; if there is not me, then everything has it. But how can a property that (negatively) tracks my existence exist in worlds in which I don't? Doesn't the property exist only if I do, to fix its application conditions?[15] More generally, (2) (even without the initial necessity operator) is hard to reconcile with the possible non-existence of any actual object, and therefore with the negation of any instance of CBF. For similar reasons, (2) (with the initial necessity operator) is hard to reconcile with the negation of any instance of BF. For if there could have been an object that actually is not, by (2) there would have necessarily been a

---

14 To check the validity of CP⁺, fix a model <W, w_0, D, dom, int> and assignment a. Let I be the intension such that for each $w \in W$, $I(w) = \{o \in dom(w): w, a[\mathbf{x}/o] \models \mathbf{A}\}$. Then for any $w \in W$, $o \in dom(w)$: $w, a[\mathbf{X}/I][\mathbf{x}/o] \models \mathbf{Xx}$ if and only if $o \in I(w)$ if and only if $w, a[\mathbf{x}/o] \models \mathbf{A}$ if and only if $w, a[\mathbf{X}/I][\mathbf{x}/o] \models \mathbf{A}$ (since **X** is not free in **A**). Therefore for any $w \in W$: $w, a[\mathbf{X}/I] \models \forall\mathbf{x}\, (\mathbf{Xx} \leftrightarrow \mathbf{A})$. Hence $w_0, a[\mathbf{X}/I] \models \Box\forall\mathbf{x}\, (\mathbf{Xx} \leftrightarrow \mathbf{A})$. Thus $w_0, a \models \exists\mathbf{X}\, \Box\forall\mathbf{x}\, (\mathbf{Xx} \leftrightarrow \mathbf{A})$, as required.

15 The objection to the idea that negative haecceities can exist in the absence of the individuals whose negative haecceities they are resembles a widespread objection to Plantinga's interpretation of quantified modal logic. Plantinga requires individual essences to exist even in the absence of the individuals whose individual essences they are. See Fine (1985) and Plantinga (1983, 1985).

negative haecceity for that object, even in the actual world, from which by hypothesis the object itself is absent.[16]

Is there a plausible metaphysics on which there can be my negative haecceity without me? That combination suggests a conception of the negative haecceity as purely qualitative. But then the nature of objects must permit them to be uniquely determined by purely qualitative properties, for negative haecceity cannot be shared.[17] For example, suppose that the same qualitative possibilities are open to Tweedledee and Tweedledum. If there had been no Tweedledee and no Tweedledee, and no other particulars specifically related to them, on this conception there would still have been their purely qualitative negative haecceities, one tracking only Tweedledee and the other tracking only Tweedledee: but how could something purely qualitative uniquely determine one of them at the expense of the other? Thus a highly contentious form of the identity of indiscernibles might well be required: no possible qualitative sameness without necessary numerical identity whenever either object exists.[18] This seems to make the nature of the objects themselves purely qualitative. But they must not be purely qualitative in the same way as the negative haecceities themselves, for the latter way of being purely qualitative is supposed to grant the negative haecceities necessary existence, whereas the opponent of BF and CBF was trying to defend the possibility of contingent existence for ordinary objects. The view might be of particulars as bundles of necessarily existing purely qualitative universals, contingently bundled together by a primitive purely qualitative higher-order multigrade compresence relation and individuated by the universals in the bundle. Such a view has no independent plausibility, and it gets worse when one tries to flesh it out with an account of the cross-world identity of bundles in a way that would avoid the Tweedledum-Tweedledee problem. All of this constitutes a heavy metaphysical burden for the opponent of BF and CBF. Although the second-order modal logic validated by

---

16 The argument at this point relies on an S5-like conception of modality: if something could have been necessary, it could have held in this actual world.

17 From $\Box\forall\mathbf{x}\,(\mathbf{Xx} \leftrightarrow \neg\mathbf{x{=}y})$ & $\Box\forall\mathbf{x}\,(\mathbf{Xx} \leftrightarrow \neg\mathbf{x{=}z})$ we easily derive $\Box\forall\mathbf{x}\,(\mathbf{x{=}y} \leftrightarrow \mathbf{x{=}z})$ from which $\Box((\mathbf{y{=}y} \lor \mathbf{z{=}z}) \leftrightarrow \mathbf{y{=}z})$ follows (since self-identity requires existence on the semantics). Thus the semantics allows no sense in which two possible individuals can share a negative haecceity.

18 Various weak forms of the identity of indiscernibles are derivable in the second-order modal logic, such as $\Box\forall\mathbf{x}\ \Box\forall\mathbf{y}\ \Box(\forall\mathbf{X}\,(\mathbf{Xx} \leftrightarrow \mathbf{Xy}) \rightarrow ((\mathbf{x{=}x} \lor \mathbf{y{=}y}) \rightarrow \mathbf{x{=}y}))$, but they are philosophically uncontentious since the relevant reading of the second-order quantifier involves no restriction to the purely qualitative (in the obvious proof, one substitutes $\mathbf{x{=}x}$ for $\mathbf{Xx}$, then $\mathbf{y{=}y}$ for $\mathbf{Xy}$).

Kripke models with variable first-order domains is formally consistent, it is philosophically very unattractive.

At this point, the opponent of BF and CBF may reflect that they had independent grounds for confining the Kripke semantics to a merely instrumental role (in what follows, talk of worlds is still occasionally used for heuristic or instrumental purposes, but plays no essential philosophical role). The fact that the semantics validates CP^+ (including all its closures) does not commit them to accepting that principle. They might limit themselves to this comprehension principle CP, a natural weakening of CP^+ obtained by dropping the intermediate necessity operator:

CP $\quad \exists X \forall x\, (Xx \leftrightarrow A)$

Like CP^+, CP counts as including all its closures. Thus the opponent of BF and CBF might still assert $\Box \exists X \forall x\, (Xx \leftrightarrow A)$, but not the result of moving the existential quantifier outside the necessity operator. The instance of CP corresponding to (2) as an instance of CP^+ is only:

(3) $\quad \Box \forall y\, \Box \exists X \forall x\, (Xx \leftrightarrow \neg x{=}y)$

For a world in which I exist, (3) yields a property that everything in that world except me has in that world, but (3) says nothing about what has that property in other worlds; in particular, it does not require the property to be my negative haecceity. For a world in which I do not exist, (3) yields a property that everything in that world has in that world, but again (3) says nothing about what has that property in other worlds; in particular, it does not require the property to be my negative haecceity – it could be the universal property of self-identity, for example.

In order to check that the move from CP^+ to CP really does have the claimed benefits for opponents of BF and CBF, we must establish that CP does not entail CP^+ or (2) by uncontentious reasoning. The notion of uncontentious reasoning here is vague, since for present purposes we can no longer rely on the Kripke semantics as a standard of validity. However, we can use the following more liberal model theory. A model is a sextuple $\langle W, w_0, D, D2, dom, int \rangle$ where all the other components are as before but D2 is a nonempty subset of the set of all 1-place intensions in the previous sense. The semantics is just as before, except that second-order variables are assigned values in D2 (although 1-place atomic predicates may be assigned values outside D2; in some sense they are not required to express properties) and the clause for the second-order quantifier is this:

$w, a \models \exists \mathbf{V}\ \mathbf{A}$ iff for some $I \in D2$: $w, a[\mathbf{V}/I] \models \mathbf{A}$

Thus D2 is the domain for the second-order quantifiers. When D2 is the set of all 1-place intensions for the model, we are back with the previous case; when D2 is a proper subset of that set, we have the modal analogue of a Henkin model for second-order logic.[19] Let us call all the sextuple models Henkin-Kripke models. For present purposes, we may reasonably assume that any argument that fails in some Henkin-Kripke model to preserve truth is contentious in the relevant sense. We do not assume that any argument that preserves truth in all Henkin-Kripke models is *un*contentious in the corresponding sense. For instance, since D2 is a constant, world-independent domain, the Henkin-Kripke semantics still validates BF2 and CBF2, but we can remain neutral for the time being on whether they are uncontentious in the relevant sense. Nor do we claim that CP itself is uncontentious; it is untrue in some Henkin-Kripke models, even when there is only one world, since the semantics does not require D2 to discriminate all subsets of the first-order domain. The opponent of BF and CBF can still defend CP and its closures without claiming them to be uncontentious. Since we can find a Henkin-Kripke model in which CP is valid but (2) is not true, we can reasonably hold that in the relevant sense CP does not entail (2).[20]

However, CP is very weak. For example, it does not entail the schema (4):

(4) $\forall \mathbf{x}\ ((\mathbf{A}\ \&\ \Diamond\neg\mathbf{A}) \rightarrow \exists \mathbf{X}\ (\mathbf{Xx}\ \&\ \Diamond\neg\mathbf{Xx}))$

A typical instance of (4) can be paraphrased as 'Whoever is contingently happy is contingently something'. That is just the kind of harmless truth that motivates second-order modal logic.[21] By contrast, (4) is true in any Henkin-Kripke model in which CP^+ is valid.[22]

---

19 See Shapiro (1991), pp. 73–76.

20 Consider the model $\langle W, w_0, D, D2, dom, int\rangle$ where $W = \{0, 1\}$, $w_0 = 0$, $D = dom(0) = dom(1) = \{2\}$, and $D2 = \{I_{yn}, I_{ny}\}$, $I_{yn}(0) = \{2\}$, $I_{yn}(1) = \{\}$, $I_{ny}(0) = \{\}$, $I_{ny}(1) = \{2\}$. CP is valid in the model, because for any assignment a: either $\{o \in dom(0): 0, a[\mathbf{x}/o] \models \mathbf{A}\} = \{2\}$, in which case $0, a[\mathbf{X}/I_{yn}] \models \forall \mathbf{x}\ (\mathbf{Xx} \leftrightarrow \mathbf{A})$, or $\{o \in dom(0): 0, a[\mathbf{x}/o] \models \mathbf{A}\} = \{\}$, in which case $0, a[\mathbf{X}/I_{ny}] \models \forall \mathbf{x}\ (\mathbf{Xx} \leftrightarrow \mathbf{A})$; either way, $0, a \models \exists \mathbf{X}\ \forall \mathbf{x}\ (\mathbf{Xx} \leftrightarrow \mathbf{A})$; similarly, $1, a \models \exists \mathbf{X}\ \forall \mathbf{x}\ (\mathbf{Xx} \leftrightarrow \mathbf{A})$; thus for any world w: $w, a \models \exists \mathbf{X}\ \forall \mathbf{x}\ (\mathbf{Xx} \leftrightarrow \mathbf{A})$. But CP^+ is invalid and (2) untrue in the model. For all first-order variables are assigned the same value, 2, which is in the domain of both worlds, so for any assignment a and $w \in W$, $\{o \in dom(w): w, a[\mathbf{x}/o] \models \neg\mathbf{x}{=}\mathbf{y}\} = \{\}$; hence if $0, a \models \Box\forall \mathbf{x}\ (\mathbf{Xx} \leftrightarrow \neg\mathbf{x}{=}\mathbf{y})$ then $a(\mathbf{X})(0) = a(\mathbf{X})(1) = \{\}$, which is impossible since $I_{yn}(0) = I_{ny}(1) = \{2\}$.

21 To show that CP does not entail (4), consider model $\langle W, w_0, D, D2, dom, int\rangle$, where W, w_0, D and dom are as in fn. 20, but $D2 = \{I_{yy}, I_{nn}\}$, $I_{yy}(0) = \{2\}$, $I_{yy}(1) = \{2\}$, $I_{nn}(0) = \{\}$, $I_{nn}(1) = \{\}$, and for

Another example of the weakness of CP is that it does not entail:

(5) $(\forall x\,(A \leftrightarrow B)\ \&\ \neg\Box\forall x\,(A \leftrightarrow B)) \rightarrow$

$$\exists X\,\exists Y\,(\forall x\,(Xx \leftrightarrow Yx)\ \&\ \neg\Box\forall x\,(Xx \leftrightarrow Yx))$$

A typical instance of (5) can be roughly paraphrased as 'If it is contingent that all and only cordates are renates, then there are contingently coextensive properties'. That is another example the kind of harmless truth that motivates second-order modal logic.[23] By contrast, (5) is true in any Henkin-Kripke model in which CP^+ is valid.

Both (4) and (5) are examples of principles that might have failed if we had read the second-order quantifiers as plural quantifiers.[24] However, some principles that do hold on the plural interpretation also cannot be derived from CP. For example, CP does not guarantee that properties are closed under conjunction. That is, we cannot derive (6) from CP:

(6) $\forall Y\,\forall Z\,\exists X\,\Box\forall x\,(Xx \leftrightarrow (Yx\ \&\ Zx))$

---

some 1-place predicate constant **F** int(**F**) = I_{yn}. CP is valid in this model, for a reason like that for the previous model. Now for any assignment a: $0, a \models Fx\ \&\ \Diamond\neg Fx$. But both $0, a[X/I_{yy}] \models Xx$ and $1, a[X/I_{yy}] \models Xx$ so not $0, a[X/I_{yy}] \models \Diamond\neg Xx$, so not $0, a[X/I_{yy}] \models Xx\ \&\ \Diamond\neg Xx$, and not $0, a[X/I_{nn}] \models Xx$ so not $0, a[X/I_{nn}] \models Xx\ \&\ \Diamond\neg Xx$. Therefore not $0, a \models \exists X\,(Xx\ \&\ \Diamond\neg Xx)$. Thus (4) is not true at 0.

22 Proof: Suppose that CP^+ is valid in $\langle W, w_0, D, D2, dom, int\rangle$. We may assume that **X** does not occur free in **A** in (4) (otherwise a slight complication in the argument is needed). Then for some $I \in D2$ and any assignment a, $w_0, a[X/I] \models \Box\forall x\,(Xx \leftrightarrow A)$, so for any $w \in W$ $\{o \in dom(w): w, a[X/I][x/o] \models Xx\} = \{o \in dom(w): w, a[X/I][x/o] \models A\} = \{o \in dom(w): w, a[x/o] \models A\}$. Suppose that $w_0, a[x/o^*] \models A\ \&\ \Diamond\neg A$ where $o^* \in w_0$. Now $\{o \in dom(w_0): w_0, a[X/I][x/o] \models Xx\} = \{o \in dom(w_0): w_0, a[x/o] \models A\}$ and $w_0, a[x/o^*] \models A$, so $w_0, a[X/I][x/o^*] \models Xx$. Moreover, for some $w \in W$ not $w, a[x/o^*] \models A$. If $o^* \in dom(w)$ then not $w, a[X/I][x/o^*] \models Xx$ because $\{o \in dom(w): w, a[X/I][x/o] \models Xx\} = \{o \in dom(w): w, a[x/o] \models A\}$. If $o^* \notin dom(w)$ then not $w, a[X/I][x/o^*] \models Xx$ because $I(w) \subseteq dom(w)$ by definition of a model. Either way, not $w, a[X/I][x/o^*] \models Xx$, so $w_0, a[X/I][x/o^*] \models \Diamond\neg Xx$. Thus $w_0, a[X/I][x/o^*] \models Xx\ \&\ \Diamond\neg Xx$, so $w_0, a[x/o^*] \models \exists X\,(Xx\ \&\ \Diamond\neg Xx)$. Thus for any $o^* \in w_0$: $w_0, a[x/o^*] \models (A\ \&\ \Diamond\neg A) \rightarrow \exists X\,(Xx\ \&\ \Diamond\neg Xx)$, so $w_0, a \models$ (4), as required.

23 The underivability of (5) can be established by exactly the same model as for (4). For $0, a \models \forall x\,(Fx \leftrightarrow x{=}x)\ \&\ \neg\Box\forall x\,(Fx \leftrightarrow x{=}x)$; yet not $0, a \models \exists X\,\exists Y\,(\forall x\,(Xx \leftrightarrow Yx)\ \&\ \neg\Box\forall x\,(Xx \leftrightarrow Yx))$ because the only two intensions in the domain of the second-order quantifiers are not coextensive at any world.

24 See Boolos (1984). Note that if 'some things' implies 'at least one thing' then a clause must be added to the interpretation to allow the value of a second-order variable to have an empty extension (as mathematical applications require).

Indeed, we can prove that CP does not entail that the second-order domain is closed under *any* Boolean combinations of **Yx** and **Zx** except for the two trivial cases of **Yx** and **Zx** themselves. For example, CP does not entail (7):

(7) $\forall \mathbf{Y}\, \exists \mathbf{X}\, \Box \forall \mathbf{x}\, (\mathbf{Xx} \leftrightarrow \neg \mathbf{Yx})$

Even when reinforced with BF and CBF, CP does not entail that the second-order domain is closed under any non-trivial Boolean combination. By contrast, (6), (7) and corresponding claims for all other Boolean combinations just are instances of CP⁺. In such respects CP⁺ yields a far more systematic theory of the second order than does CP.[25]

Quite independently of issues about contingent existence, CP is just too weak to be a satisfying comprehension principle. The opponent of BF and CBF who wishes to deny (2) needs some other diagnosis of what is wrong with it. At this point the natural suggestion for them is that cross-world variability in the first-order domain induces a corresponding cross-world variability in the second-order domain: roughly speaking, what properties there are depends on what objects there are: a property can in some sense 'involve' objects, as my negative haecceity involves me, and there could have been no property without the objects it involves. When the schematic letter '**A**' in CP⁺ is replaced by a formula with parameters, that is, free variables other than **x**, one should not

25 Consider the model in fn. 21. Let $a(\mathbf{Y}) = I_{yn}$ and $a(\mathbf{Z}) = I_{ny}$. If $w, a \models \Box\forall \mathbf{x}\, (\mathbf{Xx} \leftrightarrow (\mathbf{Yx}\, \&\, \mathbf{Zx}))$ then $a(\mathbf{X}) = I_{nn}$, but in this model $I_{nn} \notin D2$, so (6) is false at any world in this model. We can use the same model to prove that CP does not entail that the second-order domain is closed under any Boolean combinations of **Yx** and **Zx** except for ¬**Yx**, ¬**Zx** and the two trivial cases of **Yx** and **Zx** themselves. To rule out the cases of negation, consider a third Henkin-Kripke model like the previous ones except that $D2 = \{I_{yn}, I_{ny}, I_{nn}\}$. CP remains valid in this model because we have merely extended the domain of the second-order variables. But (7) is false in the model. For if $a(\mathbf{Y}) = I_{nn}$, $w, a \models \Box\forall \mathbf{x}\, (\mathbf{Xx} \leftrightarrow \neg \mathbf{Yx})$ only if $a(\mathbf{X}) = I_{yy}$, but $I_{yy} \notin D2$, so (7) is not true in this model. Thus CP does not entail that the second-order domain is closed under any non-trivial binary Boolean combination. For simplicity, the models used to establish the independence of (4)–(7) from CP had a constant first-order domain, so BF and CBF are true in these models. Thus the models also show that (4)–(7) are not derivable from CP even with the help of BF and CBF. Note that (7) is equivalent to $\forall \mathbf{Y}\, \exists \mathbf{X}\, \Box\forall \mathbf{x}\, \Box(\mathbf{Xx} \leftrightarrow \neg \mathbf{Yx})$ in Henkin-Kripke models with constant first-order domains (and no accessibility relation) but not in all those with variable domains; likewise for the corresponding closure principles for other Boolean combinations. The stronger formulation of closure is inappropriate in the current setting, since it automatically fails when the first-order domain varies, for if $a(\mathbf{x}) \notin \mathrm{dom}(w)$ then $w, a \models \neg \mathbf{Yx}$ but not $w, a \models \mathbf{Xx}$ since $I(w) \subseteq \mathrm{dom}(w)$ for any $I \in D2$. But we can also establish the independence of (4)–(7) from CP using models with varying first-order domains.

assert that the corresponding property would have existed even if the values of those variables had not.

Of course, this line of thought does not warrant a blanket ban on all instances of the comprehension schema with such parameters. Indeed, the legitimacy of such instances is essential if second-order quantification is to serve its central logical and mathematical functions. For example, consider the second-order principle of mathematical induction in a suitably extended language (where **s** expresses the successor function):

MI $\quad \forall \mathbf{X}\,((\mathbf{X0}\ \&\ \forall \mathbf{n}\,(\mathbf{Xn} \rightarrow \mathbf{Xsn})) \rightarrow \forall \mathbf{n}\,\mathbf{Xn})$

We need MI to entail consequences such as (8):

(8) $\quad \forall \mathbf{y}\,((\mathbf{Ry0}\ \&\ \forall \mathbf{n}\,(\mathbf{Ryn} \rightarrow \mathbf{Rysn})) \rightarrow \forall \mathbf{n}\,\mathbf{Ryn})$

But to derive (8) from MI in the natural way we require $\forall \mathbf{y} \exists \mathbf{X}\, \forall \mathbf{n}\,(\mathbf{Xn} \leftrightarrow \mathbf{Ryn})$, an instance of comprehension with parameters. Rather, the proposal is to permit such instances, but with a restriction to possibilities in which the values of the parameters exist. For instance, (2) might be replaced by (9):

(9) $\quad \Box \forall \mathbf{y}\, \Box(\mathbf{y}=\mathbf{y} \rightarrow \exists \mathbf{X}\, \Box(\mathbf{y}=\mathbf{y} \rightarrow \forall \mathbf{x}\,(\mathbf{Xx} \leftrightarrow \neg \mathbf{x}=\mathbf{y})))$

More generally, CP* would replace CP, where $\mathbf{v}_1, \ldots, \mathbf{v}_n$ are all the parameters in **A**:

CP* $\quad (\mathbf{v}_1=\mathbf{v}_1\ \&\ \ldots\ \&\ \mathbf{v}_n=\mathbf{v}_n) \rightarrow \exists \mathbf{X}\, \Box((\mathbf{v}_1=\mathbf{v}_1\ \&\ \ldots\ \&\ \mathbf{v}_n=\mathbf{v}_n) \rightarrow \forall \mathbf{x}\,(\mathbf{Xx} \leftrightarrow \mathbf{A}))$

An appropriate model theory would further liberalize the Henkin-Kripke semantics, by permitting variation in the second-order domain (although not all such models validate CP*). In an important respect, CP* is closer to CP^+ than to CP, because it has the crucial occurrence of $\Box$ between $\exists \mathbf{X}$ and $\forall \mathbf{x}$.

Two other variants on the same theme are:

CP*a $\quad \forall \mathbf{v}_1 \ldots \forall \mathbf{v}_n\, \exists \mathbf{X}\, \Box((\mathbf{v}_1=\mathbf{v}_1\ \&\ \ldots\ \&\ \mathbf{v}_n=\mathbf{v}_n) \rightarrow \forall \mathbf{x}\,(\mathbf{Xx} \leftrightarrow \mathbf{A}))$

CP** $\quad \exists \mathbf{X}\, \Box((\mathbf{EX} \rightarrow \forall \mathbf{x}\,(\mathbf{Xx} \leftrightarrow \mathbf{A}))\ \&\ (\mathbf{EX} \leftrightarrow (\mathbf{v}_1=\mathbf{v}_1\ \&\ \ldots\ \&\ \mathbf{v}_n=\mathbf{v}_n)))$

Under elementary assumptions, CP*a is logically equivalent to CP* (CP* has the advantage of articulating the existential condition on $\mathbf{v}_1, \ldots, \mathbf{v}_n$ uniformly in its two occurrences). In CP**, **E** is a new second-order existence predicate (questions might be raised about the intelligibility of such a predicate, but since it is built into the underlying philosophical motivation for these principles, it hardly constitutes an additional cost). CP** logically entails CP* and CP*a. The converse fails in the absence of special principles about **E** (which does not occur in CP* and CP*a). The ensuing discussion will apply to all three versions.[26]

Unfortunately for the opponent of BF and CBF, CP* is problematic in at least two ways. First, it assumes that the object-dependence of the property corresponding to **A** is exhausted by the explicit parameters in **A**. But object-dependence can arise in other ways. If it is explicit in a proper name **c** (in an extended language) rather than a variable, it can be dealt with simply by adding another conjunct **c=c** to both occurrences of the existential condition $\mathbf{v_1{=}v_1 \,\&\, ... \,\&\, v_n{=}v_n}$ in CP*.[27] But it might be purely implicit, in the intended interpretation of a non-logical atomic predicate. No obvious modification of CP* or CP*a handles this problem. We could avoid it by dropping the sufficient condition for **EX** in CP**:

CP**– $\mathbf{\exists X\, \Box(EX \rightarrow (\forall x\, (Xx \leftrightarrow A) \,\&\, v_1{=}v_1 \,\&\, ... \,\&\, v_n{=}v_n))}$

However, without a sufficient condition for **EX**, CP**– will not enable us to discharge it as an extra assumption; unlike CP**, CP**– will not entail CP*. In effect, we shall be unable to establish unconditionally any non-trivial instance of comprehension.

A second problem arises thus. In second-order logic, we can explicitly define the reflexive ancestral of a relation. Consider a two-place atomic predicate **R**. To be specific, let $\mathbf{R^{*}v_1v_2}$ abbreviate the formula $\mathbf{\forall Y\, (\forall u \forall v\, (Ruv \rightarrow (Yu \rightarrow Yv)) \rightarrow (Yv_1 \rightarrow Yv_2))}$. Then by elementary reasoning we can derive:

(10) $\mathbf{\forall X\, (\forall w \forall x\, (Rwx \rightarrow (Xw \rightarrow Xx)) \rightarrow \forall w \forall x\, (R^{*}wx \rightarrow (Xw \rightarrow Xx)))}$

If a property is inherited from parent to child, it is inherited from ancestor to descendant. But for the definition of the ancestral to have its intended effect, we should also have as a logical truth every instance of (10):

(11) $\mathbf{\forall w \forall x\, (Rwx \rightarrow (A \rightarrow A^{w}{}_{x})) \rightarrow \forall w \forall x\, (R^{*}wx \rightarrow (A \rightarrow A^{w}{}_{x}))}$

26 CP*b is the closest of the three formulas to a suggestion by Øystein Linnebo that prompted the discussion of CP*-like principles.

27 The corresponding modification of CP*a is hybrid in structure between CP* and CP*a, which is another slight reason to prefer the formulation CP*.

Here $\mathbf{A^w{}_x}$ results from substituting **x** for all free occurrences of **w** in **A**, **x** is free for **w** in **A** and does not occur free in **A**. Irrespective of the specific content of **A**, (11) should be an uncontroversial truth of second-order logic, for an object o has the ancestral of a relation R to an object o′ if and only if o′ is reachable from o by a finite number of R-steps, that is, steps in which one moves from an object to another object where the former has R to the latter. For any particular finite number *n*, we can define what it is for o′ to be reachable from o in *n* R-steps in first-order terms; if we substitute the corresponding formula for **R*wx** in (11), the result is a truth of first-order logic, irrespective of the specific content of **A**. In effect, (11) simply asserts a generalization each of whose instances is a first-order logical truth. Thus (11) should be an uncontroversial truth of second-order logic. It should be derivable from the definition of **R*** and general principles of second-order logic. A standard derivation of (11) requires an instance of comprehension for the formula **A**. The relevant comprehension principle is simply CP; the stronger principle CP^+ is not needed here.

The reasoning that supported (11) also supports its closures, such as:

(12) $\forall \mathbf{y}\, \Box \forall \mathbf{w} \forall \mathbf{x}\, (\mathbf{Rwx} \rightarrow (\mathbf{A} \rightarrow \mathbf{A^w{}_x})) \rightarrow \forall \mathbf{w} \forall \mathbf{x}\, (\mathbf{R^*wx} \rightarrow (\mathbf{A} \rightarrow \mathbf{A^w{}_x}))$

For instance, if we substitute for **R*wx** a first-order formula corresponding to the claim that one object is two R-steps from another, the result is (13):

(13) $\forall \mathbf{y}\, \Box \forall \mathbf{w} \forall \mathbf{x}\, (\mathbf{Rwx} \rightarrow (\mathbf{A} \rightarrow \mathbf{A^w{}_x})) \rightarrow \forall \mathbf{w} \forall \mathbf{x}\, (\exists \mathbf{z}(\mathbf{Rwz}\ \&\ \mathbf{Rzx}) \rightarrow (\mathbf{A} \rightarrow \mathbf{A^w{}_x}))$

This is a theorem of any standard system of first-order modal logic (without BF or CBF). The same applies to the corresponding formula for any other natural number. Thus (12) itself should be an uncontroversial truth of second-order modal logic, irrespective of the particular content of **A**. It should be derivable from the definition of **R*** and general principles of second-order modal logic. Just as (12) is a closure of (11), so the instance of comprehension needed for a standard derivation of (12) is the corresponding closure of the instance of CP needed for a standard derivation of (11):

(14) $\forall \mathbf{y}\, \Box \exists \mathbf{X} \forall \mathbf{x}\, (\mathbf{Xx} \leftrightarrow \mathbf{A})$

Now consider the case of (12) in which **◊Sxy** is substituted for **A** (think of **S** as expressing some genuine binary relation):

(15) $\forall \mathbf{y}\, \Box \forall \mathbf{w} \forall \mathbf{x}\, (\mathbf{Rwx} \rightarrow (\Diamond \mathbf{Swy} \rightarrow \Diamond \mathbf{Sxy})) \rightarrow \forall \mathbf{w} \forall \mathbf{x}\, (\mathbf{R^*wx} \rightarrow (\Diamond \mathbf{Swy} \rightarrow \Diamond \mathbf{Sxy}))$

The reason for which (12) is a truth of second-order modal logic applies in particular to (15), since it did not depend on the particular content of **A**. The instance of a closure of CP required for a standard derivation of (15) is the corresponding case of (14):

(16) $\forall \mathbf{y}\, \Box \exists \mathbf{X} \forall \mathbf{x}\, (\mathbf{Xx} \leftrightarrow \Diamond \mathbf{Sxy})$

But the metaphysical objection the opponent of BF and CBF is currently levelling against CP$^+$, the one supposed to motivate CP* instead, applies equally to (16). For (16) requires a value of **X** corresponding to $\Diamond$**Sxy** (for a fixed value of **y**) to exist even in counterfactual circumstances in which the object assigned to **y** does not. Yet (15) is a truth of second-order modal logic even without a restriction of the necessity operator by that existence condition (**y=y**). Thus the cost of the metaphysical objection to CP$^+$ is an unwarranted restriction on legitimate principles of second-order modal logic.

We cannot argue for (15) by using a value of **X** to describe from an external perspective circumstances in which that value did not exist, for the definition of the ancestral in effect quantifies only over properties that do exist in the circumstances in question.

It might be claimed that the extension of $\Diamond$**Sxy** for a fixed value of **y** in circumstances in which that value would not exist could somehow be demarcated independently of that value. But we are not looking for *ad hoc* existential claims. We want a comprehension principle that will enable us to carry out valid second-order modal reasoning. It looks as though the existential qualifications in CP* prevent it from adequately doing so.

The challenge to opponents of BF and CBF is to formulate a comprehension principle for second-order modal logic adequate by normal logical standards and compatible with a tenable metaphysics on which BF and CBF fail. If they cannot produce such a principle, the case for BF, CBF and their second-order counterparts will be very strong indeed.[28]

---

28 The argument for BF and CBF in this paper arose from further reflection on the argument for the same conclusion in Williamson (2002), which itself traces back to considerations in Prior (1967), and even further. The first step was to replace first-order quantification over propositions (such as the proposition that I do not exist) by second-order quantification into sentence position ('propositional quantification'). In place of (2), that yields something like $\Box \forall \mathbf{y}\, \Box \exists \mathbf{P}\, \Box(\mathbf{P} \leftrightarrow \neg \mathbf{y=y})$. However, the dependence of **P** on **y** required for the argument is not straightforward, for $\Box(\mathbf{P} \leftrightarrow \neg \mathbf{y=y})$ & $\Box(\mathbf{P} \leftrightarrow \neg \mathbf{z=z})$ does not entail **y=z**, only $\Box(\mathbf{y=y} \leftrightarrow \mathbf{z=z})$ (similarly, the dependence of the proposition that I do not exist on me is not straightforward on a coarse-grained conception of propositions as sets of possible worlds). The second step was

References

Barcan, R. C. [Marcus, R. B.] 1946. 'A functional calculus of first order based on strict implication', *The Journal of Symbolic Logic* 11: 1–16.

Barcan, R. C. [Marcus, R. B.] 1947. 'The identity of individuals in a strict functional calculus of second order', *The Journal of Symbolic Logic* 12: 12–15.

Boolos, G. 1984. 'To be is to be a value of a variable (or to be some values of some variables)', *The Journal of Philosophy* 81: 430–449.

Cresswell, M. 1991. 'In defence of the Barcan formula', *Logique et Analyse* 135/136: 271–282.

Fine, K. 1977. 'Postscript', in Prior and Fine 1977.

Fine, K. 1985. 'Plantinga on the reduction of possibilist discourse', in Tomberlin and van Inwagen 1985.

Kripke, S. 1963. 'Semantical considerations on modal logic', *Acta Philosophica Fennica* 16: 83–94.

Linsky, B. and Zalta, E. 1994. 'In defense of the simplest quantified modal logic', in J. Tomberlin, ed., *Philosophical Perspectives 8: Logic and Language*. Atascadero: Ridgeview.

Linsky, B., and Zalta, E. 1996. 'In defense of the contingently nonconcrete', *Philosophical Studies* 84: 283–294.

Marcus, R. B. [Barcan, R. C.] 1985/1986. 'Possibilia and possible worlds', *Grazer Philosophische Studien* 25–26: 107–133. Reprinted in Marcus 1993.

Marcus, R. B. [Barcan, R. C.] 1993. *Modalities: Philosophical Essays*, Oxford: Oxford University Press.

Muskens, R. 2007. 'Higher order modal logic', in P. Blackburn, J. van Benthem and F. Wolter, eds., *Handbook of Modal Logic*, Amsterdam: Elsevier.

Parsons, C. 1983. *Mathematics in Philosophy: Selected Essays*, Ithaca: Cornell University Press.

Parsons, T. 1995. 'Ruth Barcan Marcus and the Barcan Formula', in W. Sinnott-Armstrong, D. Raffman and N. Asher, eds., *Modality, Morality and Belief: Essays in Honor of Ruth Barcan Marcus*, Cambridge: Cambridge University Press.

Plantinga, A. 1983. 'On existentialism', *Philosophical Studies* 44: 1–20.

Plantinga, A. 1985. 'Reply to Kit Fine', in Tomberlin and van Inwagen 1985.

Prior, A. N. 1967. *Past, Present and Future*, Oxford: Clarendon Press.

Prior, A. N., and Fine, K. 1977. *Worlds, Times and Selves*. London: Duckworth.

Rumfitt, I. 2003. 'Contingent existents', *Philosophy* 78: 461–481.

Shapiro, S. 1991. *Foundations without Foundationalism: A Case for Second-Order Logic*. Oxford: Oxford University Press.

therefore to replace quantification into sentence position by quantification into monadic predicate position, as in (2), which has the further advantage that the logical and mathematical reasons for using such quantification are much stronger. The considerations in this paper can also be used against the response in Rumfitt (2003) to Williamson (2002). Williamson (2013) explores the options for opponents of BF and CBF (under the metaphysical interpretation) in more detail, and finds all of them wanting.

Stalnaker, R. 1977. 'Complex predicates', *The Monist* 60: 327–339.
Tomberlin, J., and van Inwagen, P. 1985. *Alvin Plantinga*, Dordrecht: Reidel.
Williamson, T. 1990. 'Necessary identity and necessary existence', in R. Haller and J. Brandl, eds., *Wittgenstein – Towards a Re-Evaluation: Proceedings of the 14th International Wittgenstein Symposium*, vol. 1. Vienna: Holder – Pichler – Tempsky.
Williamson, T. 1998. 'Bare possibilia', *Erkenntnis* 48: 257–273
Williamson, T. 2000a. 'Existence and contingency', *Proceedings of the Aristotelian Society* 100: 117–139.
Williamson, T. 2000b. 'The necessary framework of objects', *Topoi* 19: 201–208.
Williamson, T. 2002. 'Necessary existents', in A. O'Hear, ed., *Logic, Thought and Language*, Cambridge: Cambridge University Press.
Williamson, T. 2003. 'Everything', in J. Hawthorne and D. Zimmerman, eds., *Philosophical Perspectives 17: Language and Philosophical Linguistics*. Oxford: Blackwell.
Williamson, T. 2013. *Modal Logic as Metaphysics*. Oxford: Oxford University Press.

Pascal Engel

Is Identity a Functional Property?

Abstract

I examine the following hypothesis: identity is a second-order functional property which is multiply realisable in various domains of entities, on the model of functionalism in the philosophy of mind. If correct, such a view would enable us to solve some of the long standing difficulties of the metaphysical concept of identity, in particular the fact that our judgements of identity seem to be relative to sortals and that some of the formal features which traditionally define identity do not seem to hold universally. But the functionalist theory of identity is hardly coherent, and encounters difficulties similar to those which are raised by the parallel view of truth. I then propose to adapt a proposal by Douven and Decock who take identity judgements to be judgements of highly relevant similarity between things. This view seems to give us what a functionalist theory of identity needs. But it is an epistemological, not a metaphysical theory. It is hard to avoid the conclusion that the relation of identity has to be absolute, and cannot be reduced to any epistemological concept or to a functional property.

1 Introduction

The very first work that I read by Ruth Marcus, in 1977, was her piece "Extensionality" (Marcus 1960).[1] It awoke me from my Quinean dogmatic slum-

École des Hautes Études en Sciences Sociales (EHESS, Paris) and Université de Genève

An earlier version of this article was read at the Lauener Prize ceremony in honour of Ruth Marcus in 2008. I thank the Lauener Stiftung for this occasion to celebrate Ruth Marcus. David Wiggins and some participants convinced me that a previous approach (partially presented in Engel 2009) was hardly coherent. I am not sure that this new approach is better, but thanks to Igor Douven's and Lieven Decock's 2009 article I could attempt to formulate it. I would like to thank also Michael Frauchiger for his kindness and patience, of which I have abused. Ruth is not any more with us. I would have liked this piece to be a small testimony of my long time friendship, admiration and gratitude. I dedicate this piece to her in loving memory.

ber. It explained that extensionality is not limited to the domain of extensional logic and showed how one could construct a range of notions of extensionality of various strengths, including for intensional contexts. It also seemed to me a remarkable example of how a philosopher and logician can *construct* a concept, instead of "deconstructing" it.[2]

In the same article Ruth Marcus examined also the notion of identity and explained that the principle of the identity of indiscernibles, that if two things which have all their properties in common they are identical

(*Id ind*) $(\forall x)(\forall y)((Fy \leftrightarrow Fx) \rightarrow (x = y))$

could also be weakened (to be distinct means to be *discernibly* distinct). She also hints, at the end of her essay that when the statement "involves only proper names", then replacement of the "Morning star" by "the evening star" in such statements as

It is necessary that the evening star is the morning star

yields no paradox such as Quine's famous paradox of the number of planets. It is clearly related to her theorem about the necessity of identity, which she established in her pioneering paper of 1946, and it anticipates the famous discussion of necessary identity statements with proper names in "Modalities and Modal languages" (Marcus 1962)[3].

Traditional debates about the nature of identity bear upon whether it can really be a formal property or relation of objects, since not all of the classical logical principles of identity seem to be formal at all. This has led philosophers to propose various non classical notions of identity and in particular to relativise the concept. But these attempts meet strong objections. I want here to examine another view, which, to my knowledge, has never been explicitly proposed,

---

1 This article is not reproduced in Marcus 1993.

2 I mentioned her views on identity in a polemical piece against the atmosphere of French philosophy in the seventies *in* Engel 1978. It needed a courage that I did not have at that time to say that the Emperor of Deconstruction had no clothes. Ruth's frontal opposition to what she aptly called "intellectual fraud" was an act of intellectual responsibility, and a model for us. I find deeply comforting that Derrida's philosophy was castigated by one of the greatest logicians of our time. To paraphrase Proust quoted by Bernard Williams, it's only fifty years later that one realises how much logicians have warded off great dangers.

3 In her 1960 essay, Ruth Marcus quotes Fitch 1949, who had himself anticipated this notorious point, later, and notoriously taken up by Kripke later on.

but which is in the same spirit as the relativisation idea, since it comes down to a contextual relativisation of identity. According to that putative view, identity is a functional property, multiply realised in various domains. This proposal is modelled along the lines of the functional conception of truth which has been defended by some writers recently. The parallel proposal of a functionalisation of truth would amount to construct identity as a kind of relative similarity. I try to show that it does not succeed and that we have to stick to the classical absolute concept.

2 Absolute Identity

The difficulties of the classical definition of identity are discussed by Ruth Marcus in her essay "*Possibilia* and possible worlds" in the context of a discussion of the nature of *possibilia* (1986, 1993:189–213). Identity is taken to be a logical notion that one can characterise by the properties of reflexivity, symmetry, transitivity, indiscernibility, and substitutivity):

(R) $(\forall x)\,(x = x)$

(S) $(\forall x)\,(\forall y)\;((x = y) \rightarrow (y = x))$

(T) $(\forall x)\,(\forall y)\;(\forall z)\;(((x = y)\;\&\;(y = z)) \rightarrow (x = z))$

(*Ind Id*) $(\forall x)\,(\forall y)\,((x = y) \rightarrow (Fy \leftrightarrow Fx))$

(*Id ind*): $(\forall x)\,(\forall y)\,((Fy \leftrightarrow Fx) \rightarrow (x = y))$

It seems to be the best way to define objecthood, and a primitive notion:

> Identity is the strongest equivalence relation that a thing bears only to itself. That there are individuals is already presupposed if the identity relation is to hold. The identity relation does not *confer* thinghood; identity is an essential feature of things. Individuals must be there before they enter into any relations, even relations of self-identity. Of course if we want to *discover* which objects a language or theory takes to be individuals, we look to see which objects are such that they can meaningfully enter into the identity relation. Quantification is not so clear a guide to ontology as is identity. No identity without entity. (Marcus 1993: 200)

This reversal of the Quinean maxim is crucial. Quine was putting epistemological strictures on thinghood, by claiming, in a verificationist mood, that there is no entity without criteria of identity. Marcus says that it is thinghood which comes first. The problem, however, as Marcus reminds us, is that the notion of identity is not so clear, and that its definition in terms of indiscernibility (*Id Ind* above) – things are identical when they have all their properties in common – encounters familiar problems: what does it mean to say that two things have all their properties in common? What kinds of properties are involved (qualitative or relational, including the relation of identity itself?)? The identity of indiscernibles cannot be fully logical or formal, for indiscernibility cannot define identity without further metaphysical assumptions.

There are indeed different criteria of the formal, but on many views, identity is not a logical constant[4]. Quine could not accept the necessity of identity as a formal property, except in the *de dicto* sense. But as Ruth Marcus remarks, for instance in her "A Backward look at Quine's animadversions on modalities" (Marcus 1988; 1993: 229) and in "Essentialism in modal logic" (Marcus 1967; 1993: 50), the necessity of identity does not need other kinds of essential properties than self-identity, and is committed only to an innocuous kind of essentialism. So there *is* a good ground to accept the necessity of identity among the formal-logical properties of identity. The situation is less clear with the non vagueness of identity, which has been defended by Gareth Evans on the basis of a famous argument which is reminiscent of Marcus's proof of the necessity of identity (Evans 1982)[5]. A lot of discussion has been devoted to whether identity *statements* can be vague or not, but it seems difficult to deny that the property of an object to be self-identical could be vague. So there is good ground for

---

4 On Peacocke's criterion for logical constanthood (Peacocke 1976), identity does not come out as a logical constant.

5 (1) ∇ (a = b) (where '∇' is the operator 'vaguely')
(2) ŷ [∇ (y = a)] b
(3) ~∇ (a = a)
(4) ~ ŷ [∇ (y = a)] a.
From Leibniz's law of substitutivity one can derive from (2) and (4):
(5) ~ (a = b)
Which contradicts the initial hypothesis that the identity statement 'a = b' is indeterminately true. On Evans' argument, see Williamson 1994: 253–56, and the references thereof. The proof given independently by Salmon 1977 is actually simpler. See also Salmon 2002. (1) Suppose a pair, <x, y>, for which there is no fact of identity or distinctness. (2) By contrast, there is a fact of identity for the reflexive pair <x, x>. (3) It follows that <x, y> is distinct from <x, y> (4) Therefore, by standard set theory, x ≠ y. (5) Consequently, there is a fact of the matter. See also Engel 2003, Parsons 2000.

adding to the classical list of formal features of identity (reflexivity, transitivity, indiscernibility, substitutivity) those of necessity and non-vagueness. All these make identity an *absolute* relation, which holds of objects independently of what kind of objects enter in the relation. In other words absolute identity characterises identity as a formal property in the sense that it is supposed to apply to any kind of object, whether it be physical, biological, mental, natural, artefactual, etc. Any two objects, if they are identical, are identical *simpliciter*, and not relative to a kind or sort.

In the same article (Marcus 1993: 228) Ruth Marcus notices that for Aristotle, essential properties are sortal, in contrast to the property of *being an entity* and the property of *being self-identical*. It is precisely because they wanted to include sortal properties within the property of *being identical* that some philosophers have attempted to *relativise* identity itself to sortals. Geach (1972) claimed that the absolute concept of identity does not fit our usual way of asking identity questions. When we ask whether one object is identical to another, we never ask whether they are identical *simpliciter*, but we always ask questions of the form: "What kind of thing is x?" and we always reply, in the Aristotelian mode, that x is an F, or G, etc. When we ask whether x is identical to y, it is always relative to a sortal F, G, etc. This is, in a sense, how it should be, since the formal properties of identity (reflexivity, etc.) do not determine what kind of thing x and y are, or under what respect x and y are identical. In traditional terminology, what determines which kind of thing is x is what individuates x, and individuation is another matter than identity.

Actually more or less all the kinds of problems which are called "problems of identity", those which give rise to famous paradoxes (material composition *vs* form, identity over time, Theseus' ship, the problem of the many, etc.) are problems relating to individuation. Take for instance Geach's case of the cat Tibbles as an illustration of what is called "the problem of the many".[6] Tibbles is a cat. Let us suppose that he loses his tail, the resulting object being Tib. Tib differs from Tibbles. But Tibbles minus his tail coincides with Tib. And if Tibbles is a cat, Tib is a cat too. One can say that Tib differs from Tibbles as an individual but is identical to Tibbles as a cat. Aren't there two Tibbles? A variant has Tibbles losing his hair on the carpet: if Tibbles loses 1000 hairs, 1000 Tibbles are on the carpet.

---

6 A paradox, which, as Wiggins 1980/2001, pp. 173–76 notes, comes from Chrysippus. The phrase comes from Peter Unger 1980.

According to Geach, most of these paradoxes show that the predicate of identity, if it is to make sense within ordinary contexts, has to be relativised. "$a = b$" has always to be understood as

(R) *a* is the same F as *b*

where 'F' is a sortal predicate. The problem, as David Wiggins has shown, is that relative identity (R) is incompatible with the principle of the identity of indiscernibles [7]. The dilemma is: either identity is absolute, but it can't really be a relation in which ordinary objects enter, or it is relative, but then it ceases to be a genuinely formal relation.

Geach's problem about relative identity and the issues raised by the paradoxes of identity show that there is a sharp contrast between what logic tells us about the identity relation – which is absolute, simple, and necessary – and our *identity judgments* – which are relative, context-sensitive and contingent. Geach is right that in many contexts we do not ask whether a thing is identical to another *simpliciter*. Very often we satisfy ourselves with vague or loose identity statements ("This is the same food as the one I had in another restaurant") or ambiguous ones (for instance we are happy to accept certain type identity statements although we would reject their token counterparts: "This is the same ship as the one I took for Chios last year"), and we rarely care for what Thomas Reid called the "strict and philosophical sense" of identity (we almost never ask: "Is it *really* the same food as the one from the other restaurant ?", and when we ask "Is it the *very same* ship as the one you took for Chios last year?" we pose a quite different kind of question from the one quoted above). In such cases, identity is neither absolute nor transitive, and seems akin to a notion of similarity. Indeed there are cases where we are interested in the strict sense, as when an Egyptologist asks whether the statue that he inspects in an antiquarian shop is the same as the one stolen last month by looters from excavations near Alexandria. But can we content ourselves with the observation that there is a conflict between the loose and ordinary sense on the one hand and the strict and philosophical sense on the other?

Although the notion of identity itself is not *prima facie* epistemic, the notion of an identity *judgement* is largely epistemic, and for this reason it is person and context relative. What counts as identity for one person in one circumstance depends on her interests and on her epistemic situation. The difference between

7 Cf. Wiggins 1980/2001, *op.cit*, ch.1, Engel 1989 p. 241.

identity judgements from one context to another are obviously differences about the epistemic *criteria* by which we judge one thing to be identical to or to differ from another, or differences about our epistemic interests in asking identity questions. Identity questions, however, are not only epistemological questions: they are also, and quite often, questions about what things are and about their nature. Now, even if we understand these questions as ontological questions, the problem of the variety of identity questions still arises. When we ask whether the statue that we are looking at in a shop in the Plaka is the same as the one in the Acropolis museum, we wonder about a difference in *constitution*. When we ask whether Theseus' ship at one time is the same as a ship of the same shape at another time, we wonder about a difference in material *composition*. These are principles about the individuation of entities in an objective and not in a mind dependent sense. But these questions about individuation differ from identity questions "strictly" speaking. But we cannot rest content with the thought that questions about identity are distinct from questions about individuation. We want to know how they are related. Does individuation depend on identity or *vice versa*?

3 Wiggins' Individuative Essentialism

David Wiggins (1980/2001) proposes a solution to this problem. He holds that any entity has to conform to the formal principles of the absolute concept, which are norms for specific issues about individuation which occur in various contexts and relative to various purposes that we have of classifying objects into kinds. Without these principles, no judgement to the effect that x is identical to y can make sense. According to Wiggins, however, it does not follow from this that attributions of identity are independent of the implicit reference to *some* sortal or other. Although identity is absolute, "the sortal dependency of individuation" still holds. According to Wiggins' "individuative essentialism" it has to be true both that:

1. Identity is absolute and defined by its formal properties (*R*, *S*, *T*, *Ind Id*, *Id Ind*)

2. A thing x is identical to another y if and only if there is a sortal concept F such that x and y fall under F

> 3. This concept provides the individuation of the entities in question and is a principle of *coincidence* and of *activity*, of *functioning* and *operation* (in other words it is a substance concept) (Wiggins 2001: 72–73)

According to individuative essentialism, individuation is relative to a common sortal concept which underlies any individuation, but this concept is left unspecified by the schema of the theory. It may be any sortal concept which is general enough to track the object. But this sortal concept does not vary from context to context, from one identity question to another. On the contrary the same sortal is presupposed by any individuation. It is the concept of what Wiggins calls a "continuant".

Such a view seems equipped to give a solution to some of the paradoxes about identity, such as the problem of the many. Confronted with the Tib-Tibbles case, individuative essentialism will say that Tib and Tibbles are two distinct entities which coincide under the same sortal (*cat*). Wiggins rejects any view which would say that Tib and Tibbles are composed of distinct spatio-temporal parts. Their difference is based upon their constitution. Although identity is a stable property, there are variations in the conditions of persistence, which differ according to whether one deals with material objects, biological organisms, persons, artefacts or different kinds of entities. Each one of these has its own individuative kind, but continuants falling under these kinds have distinct persistence conditions. Constitution does not *define* identity, but identity supervenes on constitution. Identity is any relation which conforms to the formal principles (*R*), (*S*), (*T*), (*Id Ind*), (*Ind Id*), but for each kind of entity, there is a substantive account of the individuation of members of that kind, and conformity to the formal principles is a consequence of that account.

But can Wiggins' individuative essentialism solve our problem? If persistence conditions vary with the kind of entity considered, and if identity supervenes on persistence conditions, how can we escape the conclusion that identity is nothing over and above the constitution of the kinds? Do the formal principles hold in the same way in all domains? For instance if we accept that the constitutive principles for personal identity are based on a notion of psychological continuity of the Lockean or Parfitian kind, will identity be transitive[8]? Can't we accept that the principles of individuation of some entities, such as social groups like clubs or associations, can be vague (a club does not cease to exist if it loses one of its members, but a loss of ten members may be fatal,

8 See Parfit 1984, Wiggins discusses these issues in chapter VII of Wiggins 2001.

where does the limit fall?). Do the formal principles of identity hold in basics physics? A number of writers hold the view that in quantum physics the identity of indiscernibles fails as well as the principle of the non-vagueness of identity.[9] The same is true of biological entities.[10] The individuation conditions of an organism depend upon persistence conditions which differ from those of physical objects, artefacts or persons. These conditions are those which guarantee the continued process of the life of the organism. But things are not so clear at the beginning of life: for instance the possibility of monozygotic twins suggests that a zygote is not a single and coherent life, but a system which has the potential capacity to form several lives. Hence in what sense are we confronted with a coherent whole or with a set of autonomous cells? The individuation conditions are not clear either at the end of the life of (human) organisms: medical practice has replaced cardio-pulmonary death with cerebral death. Should we then say that the irreversible loss of cerebral functions constitutes the death of the person or that of the organism? Another example of the complexity of identity questions for living beings is the nature of the "self" in immunology. According to Carosella and Pradeu (2006) the classical model in immunology which rests upon the distinction between the self and the non self presupposes that every element external to the organism provokes a reaction designed to preserve its integrity. But immunitary reactions are not limited to exogenic factors (bacteries, parasites, foetuses are tolerated by the self). This is why, according to Carosella and Pradeu, the criterion of identity is continuity rather than identity. They show that any strong discontinuity within immune receptors and their targets gives rise to a reaction of the immune system, which does not discriminate between self and non-self, but between endogenous or exogenous epitopes which are constantly present and others which break the continuity of interactions. This is why, according to them, biological identity is nothing but a kind of continuity:

> The continuity hypothesis conceives of identity as an identity-continuity, since it claims that nothing more than the spatiotemporal continuity of adhesions between immune receptors and ligands defines immune identity. This hypothesis can therefore be seen as the immunological point of view on the identity of organisms. According to the continuity hypothesis, nothing like a permanent 'core' to be preserved against all foreign threats is presupposed and thought to define immunity. Changes from the inside and changes from

---

9 See, within a large literature devoted to this topic, S. French and M. Redhead 1988 for the identity of indiscernibles, and Lowe 1994 for the principle of the absoluteness of identity and how far it stands in quantum physics.

10 See in particular the interesting article by Boniolo and Carrara 2004.

> the outside equally can trigger an immune response, depending on the conditions of encounter. (Carosella and Pradeu 2006: 247)

Another problem raised by Wiggins' conception is that it seems to presuppose a form of Aristotelian essentialism in biology. He holds in particular that the sortals upon which continuants depend are natural kinds. A number of writers, however, starting with David Hull, have held that biological entities such as species can be considered as individuals and not as kinds. If this is correct one would need a different conception of the sortals upon which the continuants depend.

Most of the writers who criticize the use of the classical and absolute notion of identity suggest that we should dispense with these principles and replace them with principles for a weaker relation than identity, namely continuity, for which not all of the principles such as that of the identity of indiscernibles are valid. The view that identity principles can be in some sense reduced to the principles which hold for constitution (mostly distinct continuity principles for the various kinds of entities) raises, however, serious problems. The first one is that it resurrects the ghost of the thesis of the relativity of identity, and of a pluralistic view according to which there are as many types of identity relations as there are principles of individuation in each domain. Thus there would be physical identity, biological identity, personal identity, social identity, etc. Some writers, as Boniolo and Carrara 2004, simply accept this form of relativism. But then it is hard to see how it can escape the difficulties adduced for Geach's view. Or one has simply to bite the bullet, and reject the classical and absolute notion of identity.

The second difficulty of the pluralistic or anti-essentialist conception consists in its reduction of identity to continuity. As many discussions on the problem of personal identity have shown since Locke and Butler, identity is not the same thing as continuity.[11]

Our problem is this: how can we conciliate the Leibnizian absoluteness and the sortal dependency of identity? On the one hand, on Wiggins's view, any relation which is a genuine relation of identity must conform to the Leibnizian schema

(*Ind Id*) $(\forall x)(\forall y)((x = y) \rightarrow (Fy \leftrightarrow Fx))$

---

11 See the classical objection to Locke by Joseph Butler, *First Dissertation to the Analogy of Religion*, 1736, repr. in Perry 1977. For some of the reasons to stick to Butler's view, see Wiggins 2001, ch. VII.

but for each kind of entity, there is a substantive account of the individuation conditions of that kind, and conformity to the schema is entailed by this account. The pressure, however, for this view, comes from the fact that the individuation conditions do not in all cases entail the Leibnizian principle. On the other hand, if we reduce identity to continuity, it is at the price of losing the formal nature of identity and its absoluteness. Someone could say at this stage: so much the worse for the formal character of identity and its absoluteness. But then how can we say that we have a theory of identity?

I shall, at the end of the day, side with a conception of identity which comes close to Wiggins' and Marcus' absolutism. But in order to take the foregoing objections seriously and to try to keep in the spirit of Wiggins' conciliation, we may try to formulate another conception, which, like Wiggins' attempts at a conciliation of Leibnizian absoluteness with the variability of our identity judgements.

4 Functionalism About Truth and Identity

How can we reconcile the unity and the absoluteness of the identity relation with the plurality of the modes of constitution and of the conditions of persistence of objects of different kinds? A possible solution takes its inspiration from the familiar functionalist conception of mental states in the philosophy of mind. According to functionalism, the mental properties of an organism are defined by the causal role of the properties of this organism. In most contemporary versions of the view functional properties are second-order properties: they are role-properties of first-order (physical and biological) properties of organisms. These roles are "multiply realised" in these first-order properties of the organisms, and supervene upon them, without being identical to these properties.

This familiar functionalist picture can be extended to other kinds of properties of a more formal or abstract sort. In particular a number of writers have proposed that *truth* could be conceived as a functional property defined at the abstract level as satisfying a set of "platitudes", and realised in various domains[12]. Just as, according to functionalism about mental states, our use of "belief", "desire" and other mental terms, can be individuated in terms of the role that they play, together with other states, in mediating between inputs and

12 Pettit 1996, Wright 2002, Lynch 2009. I have myself suggested the view in Engel 2002.

outputs, we could say that the predicate “true” is a place mark for a certain role marked by a set of platitudes which Crispin Wright (2001: 760) lists thus:

- *transparency*: to assert (believe) that *p* is to present *p* as true
- *epistemic opacity*: some truths may not be known or be unknowable
- *embedding*: truth aptness is preserved under various syntactic operations
- *correspondence*: for a proposition to be true is to correspond to reality
- *contrast*: a proposition may be true without being justified and vice versa
- *stability*: if a proposition is ever true, then it is always true
- *absoluteness*: truth is absolute, there are no degrees of truth

Just as one can, for the property of pain functionalise this property by conjoining all the properties characteristic of its functional role by forming the corresponding familiar device of “ramsification”[13], we can functionalise truth:

> **FT** There is a property T which plays the truth role if and only if P is T iff P & P is T iff P is such that things are as P says they are & P is T if it is such that P can be justified but not T & P is T iff P is stable & T is the norm for belief, etc.

The issue of the nature of the properties which “realise” these roles is left open. Truth is a property which can be variably realised, just as a functional property can be so. Summarizing the idea, Crispin Wright says:

> The concept of truth admits a uniform characterisation wherever it is applied – the characterisation given by the minimal platitudes, which determine what is *essential* to truth ... The form of pluralism for which space is allowed by this overarching uniformity is *variable realisation*. What constitutes the existence of a number is different from what constitutes the existence of a material object. (Wright 1996: 924)

In other words truth is a second-order property of our statements, which has to be realised in various ways in first-order properties which will underlie this role.

But this view is unstable[14]. In Lynch’s (2009) version of alethic functionalism truth is the second-order functional property of having a property which plays the “truth role”, relative to a given domain. But is truth the higher-order property or is it the disjunction of the realiser properties? But we cannot identify truth with the realiser properties, because, given that these are by definition

---

13 See Lewis 1966 for the Ramsification technique.

14 As it has been argued by N.J. Pedersen 2010 and C. D. Wright 2010 in particular.

distinct, what is common to them would be lost, just as the common explanatory power of truth would itself be lost. But if truth is in one domain correspondence, in another superassertiblity, yet in another one coherence, it becomes unclear what common property these realise (Lynch 2009: 66). For this reason Lynch prefers to say that:

> Truth is, as it were, *immanent* in ontologically distinct properties. Let us say that where property F is immanent in or *manifested by* property M, it is *a priori* that F's essential features are a subset of M's features ... Propositions about different subjects can be made true by distinct properties each of which plays the truth-role. Thus (atomic) propositions about the antics of the ordinary objects and properties of our daily life may be true because they represent those objects and properties. For propositions of that kind, correct representation plays the truth-role and it is a priori that if a proposition correctly represents it will be true. For propositions of another sort, perhaps moral propositions, superwarrant may be what plays the truth-role, or manifests truth. (Lynch 2009: 74, 77)

But if we pause for a minute to think about what this entails, it is by no means obvious that the view is coherent. Will the truish features which are essential to the property of truth be the same in all domains? No, since by definition, alethic functionalism entails that the properties which realise truth will not be uniform in all domains. In ethics truth may be realised as coherence, whereas it can be realised as correspondence in, say, physics, or as superwarrant[15] in mathematics, etc. For the view to make sense, the realiser properties must be such that the properties which play the truth role are uniform. But how can they be uniform if they are realised in different ways in each domain? The only way to make them uniform is to limit the truth role to the most trivial properties, such as the syntactic features of the truth predicate and the deflationary schema *'P' is true iff P*. But then the view becomes hard to distinguish from deflationism about truth, the view that truth is but a "thin" property whose "essence" is exhausted by the deflationary schema. This is an unwelcome consequence, because functionalism about truth was supposed to give us an alternative to deflationism.

I shall not here dwell upon the difficulties of alethic functionalism[16]. My purpose has been to introduce it in order to see whether a similar idea can be exploited with another "formal" notion, the notion of identity. Actually the functionalist move has a long ancestry in scholastic metaphysics with other "formal" or what were called "transcendental" notions (*res*, *verum*, *aliquid*,

---

15 Superwarrant is the idea, borrowed by Lynch to Crispin Wright, that a statement has been justified to the extent that it has survived all attempts to refute it.
16 See Engel 2013.

bonum, unum). Take the predicate "exists"[17]. It is one thing for a sensation to exist, another thing for an image, yet another thing for a tree or a person, etc. Should we say that the notion of existence is ambiguous? This was the traditional view of the analogy of being. But we also want to say that there is something common to these different senses of "exist", a property which applies to all sorts of entities but which is realised in different ways whether it applies to numbers, material objects, persons or images. Should we take the functionalist move for identity? Let us see what it would look like.

Identity is the one and only relation captured by the "platitudinous" principles of reflexivity, symmetry, transitivity, substitutivity and the identity of indiscernibles, but there are varying conditions of identity for material objects, persons, events, etc. So identity is a multiply realisable role-property of the second-order.

> **FI** There is a property I that plays the identity role iff for all x, x I x, & for all x,y,z if x I y and y I z, then x I z & if two things are I then they have all their properties in common, etc.

The idea is that this functional property will be multiply realised depending on the kinds of objects to which it applies: the property of identity will be different depending upon whether we deal with material objects, artefacts, persons, or events, etc. In each domain there will be distinct individuation conditions. As far as I know, no one has proposed this kind of view in the literature. But it seems at least to make sense.

The problems, however, encountered with the functionalisation of truth reproduce with FI. If identity is the functional role made up of the conjunction of these "platitudes", should we say that identity is among the realiser properties – identity for artefacts, identity for persons, identity for material objects, etc. – or at the level of the higher-order property? If the former, identity, as we saw for biological objects, can fail to realise some of the platitudes, such as transitivity, and the functional role ceases to be common to all the realisers. If the latter, the notion of identity becomes so thin and abstract that it cannot take in charge the conditions of individuation. We have made little progress. And the same instability as the one that affects truth functionalism holds for identity functionalism: either identity is uniform across the domains in which it is realised, but then it cannot be plural, or it is plural, but it looses its uniformity. According to

17 To my knowledge, the first to note the analogy between "exists" and "is true" in this con text is Mark Sainsbury in his comments on Wright 1992, see Sainsbury 1996.

identity functionalism, several ways in which things can be identical, and for a given identity relation *I*, the way in which *I* holds depends on its specific domain. For instance Tibbles is the same as Tib *qua* animal but not *qua* material object.

Suppose that several ways of being identical are $I_1 \ldots I_n$. Given $I_1 \ldots I_n$, one can introduce a new way of being identical, IU. When are two things *x* and *y* IU? They are IU if and only if they are I_1 or ... or I_n. In other words

$$\text{IU} \quad (\forall p)\,(\text{IU}(x, y) \leftrightarrow I_1(x,y) \vee \ldots \vee I_n(x,y))$$

But the characterization of IU makes being identical in one of the ways $I_1 \ldots I_n$ a necessary and sufficient condition for being IU. That is, the things which are identical in the IU kind of way are *exactly* those things which are identical in one of the ways endorsed by the functionalist. This means that IU is a universal and uniform way of being identical: any statement whatsoever is true if and only if it is IU. The pluralist is thus committed to IU. But IU is what the functionalist denies[18].

5 Functionalism About Highly Relevant Similarity

Clearly what is wrong with functionalism both about truth and about identity is that the list of the second-order functional properties which characterise the truth role or the identity role are far too general and abstract to be realised in the low level properties to meet the plurality of ways of being identical. It is the very idea of there being many ways of being identical or of instantiating the identity relation which is problematic, and here the syndrome is the same as with relative identity. What we need then are weaker higher-order properties. In the case of identity which interests us here, the problem lies in the fact that identity judgments do not conform to the formal properties of identity: (i) they are, as Geach remarked, very context sensitive in ordinary life, (ii) they do not respect, as the case of biological entities and perhaps persons shows, the principle of transitivity, and (iii) the identity of indiscernibles and perhaps the absoluteness of identity do not seem to apply in quantum physics. One familiar strategy, if one wants to keep with the idea that the concept of identity is a formal concept, consists in accepting different logical principles depending on the

---

18 I reproduce here Pedersen's 2010 formulation.

domain. For instance, one could adopt mereology for issues about constitution of material objects, or one could reject the transitivity of continuity for personal identity. The problem, as with all uses of non-classical logic for dealing with truth value gaps or other anomalies, is that this strategy is not uniform and has a case to case flavor. In this respect the functionalist idea preserves the unity of the higher-order property.

The alternative strategy consists in taking as primitive not the classical notion of identity, but a weaker notion, close to identity, such as *relevant similarity*. It has been often argued (e.g. by Jubien 1997) that most of our identity judgments are actually judgments about relevant similarity. This is what Douven and Decock (2010) propose. They argue that many of our ordinary judgments about identity are not really about whether things are identical in the absolute sense, but are about whether things are highly similar in all relevant respects, depending on the context at hand. In other terms, on their proposal:

> An ordinary identity statement "a is identical to b," made or evaluated in a conversational context C, is to be understood as a claim to the effect that a is identical C to b – or Id C(a,b), for short – which in turn is taken to mean that a is highly C similar to b in all relevant C respects. In other terms, Id C(a,b) iff for every relevant C respect r, any difference in r between a and b is, if it exists at all, negligible C, where it is assumed that a difference that is negligible in one context need not be so in another. (Douven and Decock 2010: 67)

It looks, prima facie, as though it were a version of the thesis (R) above, of the relativity of identity. But, as they point out, Douven and Decock do not relativise identity to sortals, but to contexts, the point being that to each context corresponds a certain kind of similarity. To each context C is paired a metric similarity space SC appropriate to that context[19]. For each respect that is relevant in a context, the context contains a corresponding similarity space. For instance if colour is a relevant respect in C, then SC will contain a colour space; if shape is relevant, SC will include a shape space; if time is relevant, it will include a temporal space; and so on. For each similarity space $r \in SC$; let $d_r(.,.)$ be the distance function associated with that space; so $d_r(a,b)$ measures the distance between the representations of objects a and b in r. Also associated with each $r \in SC$ is a threshold value $t_r C > 0$; which may be different for different r and also – though for reasons given above this may be taken as optional – different for the same r in different contexts C. Then the ordinary or folk concept of identity is defined thus:

19 The notion of similarity space is familiar from recent work in cognitive science, in particular Gärdenfors 2000.

$$I\,d_C\,(a, b) \Leftrightarrow (\forall r) \in SC:\; d_r\,(a, b) \leq t^r{}_C$$ (Douven & Decock 2010: 67)

What counts as highly similar varies from context to context. This explains why in some cases transitivity of high similarity does not hold. They point out, however, that there is no reason to doubt that in most cases our ordinary folk identity judgements respect symmetry and reflexivity. The contextual similarity approach also allows a version of the indiscernibility principle. A property is a region in a similarity space; Douven and Decok show that although the revised version of (*Id Ind*) does not entail the sharing of all properties for identity, it entails the near sharing of all relevant properties (Douven and Decock 2010: 68–69).

Although it is distinct from the notion of relative identity one might worry that the relation of high similarity in relevant respects *Id C* (x,y) may suffer from the same difficulties as those of (R). In particular it seems that we can reproduce the reasoning which led Wiggins to conclude that relative identity is incoherent, if we accept that two things can be the same F but not the same G, and transpose to contexts[20]. But on reflexion, this should not be a problem: something can very well be Id (C) to another in context C while not being Id (C*) in another context C*. And the indiscernibility of identicals does not hold without restrictions. It is possible for there to be indiscernible objects which are distinct and distinct objects which are indiscernible, given the appropriate contextual restrictions.

It seems that the relation Id (C) is the appropriate candidate for a functionalist theory of identity. The relation of highly relevant similarity is the one which plays the identity role:

---

20 Given Leibniz's law in the IdC version: $x\ Id\ y \rightarrow (\phi x \leftrightarrow \psi y)$,
and given by (Id C) where C is a context: $x\ Id\ y \leftrightarrow (\exists C)\,(x\ Id_C\ y)$
we obtain $(\exists C)\,(x\ Id_C\ y) \rightarrow (\phi x \leftrightarrow \psi y)$
and we choose a context D such that $(\forall D)\,(x\ Id_D\ y \rightarrow (\phi x \leftrightarrow \psi y))$
by definition on the Id (C) relation, and for an arbitrary context E
we have $(x\ Id_D\ y \rightarrow (Ex \rightarrow x\ Id_E\ y))$.
Now if x depends on context E, we have $(x\ Id_D\ y\ \&\ Ex)$.
But then $x =_E x$. By Leibniz's law if there is predicate $x\ Id_E \ldots$, y has this predicate.
So we have $x\ Id_D\ y \rightarrow (Ex \rightarrow x\ Id_E\ y)$.
But this entails the denial of Id C:
$(\forall E)\,(\forall D)\,(x\ Id_D\ y \rightarrow (Ex \rightarrow x\ Id_E\ y))$
I have only transposed to contexts the proof given by Wiggins 2001: 53–54. Since the high similarity proposal does not accept Id C $x\ Id\ y \rightarrow (\phi x \leftrightarrow \psi y)$ unrestrictedly, the proof is blocked. Douven and Decock also show that their view is not subject to the kind of objection that Koslicki 2005 pressed against indeterminate identity views.

> **F Id (C)** There is a property Id (C) that plays the high similarity role iff for all x, x Id (C) x, & for all x,y,z if x I (C) dy and y I z, then x I z & if two things are Id (C) then they have *nearly* all their properties in common ...

This role is the set of features which are uniform across the various kinds of context in which the Id (C) relations hold. But the relation is to be distinct in different contexts. It can be realised in one way for certain entities, in another way for other entities. Each realiser corresponds to a distinctive similarity space and a value within that space. But it is not, unlike for truth and identity, implied that the property Id (C) is uniform across domains. So we do not have the problem of instability encountered in the previous functionalisation of truth and identity.

The main objection to the functionalisation of truth FT was that the status of the property of truth is uncertain: it appears both at the functional higher-order level and at the realiser level. Here with FI we do not encounter this problem, for it is not identity which appears at the higher-order level, but highly relevant similarity. Identity may appear at the realiser level, when the kind of question which is being asked is a question of identity, in the strict sense.

Unfortunately, F Id (C) cannot give us what we need, and for a very simple reason. It is not an account of the functional property of *identity*, but an account of the functional *concept* of similarity. As Douven and Decock make it clear the relation of highly relevant similarity is an *epistemic* notion, which applies to epistemic judgements of identity and their contextual relativity. A functionalist theory, however, is not a theory about our concepts, but a metaphysical theory about the nature of *properties*. It is properties which have realisers, not concepts. To concepts correspond different kinds of judgments, not entities of a given sort. I fully agree with Douven and Decock that Id (C) captures well our epistemic judgments of identity. But we cannot, on pain of begging the question, assimilate that relation to identity. Douven and Decock tell us that Id (C) captures the folk notion, or the "loose" sense of identity, but not what Reid called identity in the "strict" metaphysical sense. Indeed one might answer that I am myself begging the question in supposing that the metaphysical sense of identity is distinct from the epistemological one. But unless one has shown that the latter can be reduced to the former, the absolutist sense of identity is to stay with us, for we want to be able to ask whether two *things* are identical or not, *simpliciter*. In order to deal with the ontological issue and to find an equivalent of the functionalist view of identity for properties, one would have to defend a notion of weak identity and of weak discernibility. There are attempts in this direction, including defences of vague identity against the absolute notion

(Parsons 2000). But it is not clear that they succeed (Salmon 2002, Hawley 2009). I shall not examine this point here. The point, however, that I want to press is that if a functionalist theory of identity can work, it would have to have a *uniform* functional property, which would be along the lines of FI. And if it is along these lines, it is bound to be unstable, as I have argued above.

6 Conclusion

I have tried to make sense of one alternative to the classical absolute notion of identity, which I have formulated on the model of the functionalist conception of truth. Both are unstable, and cannot locate the functional property at the correct level. The notion of highly relevant similarity seems to give us what we need, but it is a purely epistemological theory. So if we want to have a theory of the metaphysical relation of identity, we have to stick to the absolutist view, the one that Ruth Marcus showed to be dependent on the very notion of an object[21].

References

Boniolo, G. and Carrara, M. 2004 "On Biological Identity", *Biology and Philosophy* 19: 443–457.
Carosella, E. and Pradeu, T. 2006 "The Self model and the conception of biological identity in immunology", *Biology and Philosophy* 21: 235–252.
Douven, I. and Decock, L. 2010 "Identity and Similarity", *Philosophical Studies* 151, 1: 59–78.
Engel, P. 1978 "Du bon usage des banalités", *Critique* 369: 165–177.
Engel, P. 1989, *La norme du vrai*, Paris, Gallimard, engl tr. *The norm of Truth*, Harvester, Hemel Hemstead, 1991.
Engel, P. 2002 *Truth*, Chesham, Acumen.
Engel, P. 2003 "Les objets vagues le sont-ils vraiment?", *Cahiers de Philosophie de l'université de Caen* 40–41: 103–20.
Engel, P. 2009 "L'unité de l'identité et la pluralité des individuations", in Carosella, E. ed. *L'identité*, Paris, Harmattan.
Engel, P. 2013 "Alethic Functionalism and the Norm of Belief", in N.J. Pedersen and C.D. Wright, eds., *Truth and Pluralism: Current Debates*, New York, Oxford University Press, 69–86.
Evans, G. 1982 "Vague identity", *Analysis* 3: 128–30.
Fitch, F. 1949 "The Problem of the Morning Star and the Evening Star", *Philosophy of Science* XVI: 137–41.

21 Ruth Marcus "Possibilia and Possible Worlds", in Marcus 1993, p.213.

French, S. and Redhead, M. 1988 "Quantum Physics and the Identity of Indiscernibles", *British Journal for the Philosophy of Science* 39: 233–46.
Gärdenfors, P. 2000 *Conceptual Spaces*, Cambridge Mass., MIT Press.
Geach, P.T. 1972 *Logic Matters*, Oxford, Blackwell.
Hawley, C. 2009 "Identity and indiscernibility", *Mind* 118: 101–19.
Jubien, M. 1997 *Metaphysics*, Oxford, Blackwell.
Koslicki, C. 2005 "Almost identical objects and the suspect strategy", *Journal of Philosophy* 102, 2: 55–77.
Lewis, D. 1966 "How to define Theoretical terms", repr. in *Philosophical Papers*, vol. 1, Oxford, Oxford University Press, 1986, 78–95.
Lowe, E. J. 1994 "Vague Identity and Quantum Indeterminacy", *Analysis* 54: 110–14.
Lynch, M. 2009 *Truth as one and many*, Oxford, Oxford University Press.
Marcus, R.M.B. 1946 "A Functional Calculus of First Order Based on Strict Implication", *Journal of Symbolic Logic* 11: 115–18.
Marcus, R.M.B. 1947 "The identity of individuals in a strict functional calculus of first-order", *Journal of Symbolic Logic* XII: 12–15.
Marcus, R.M.B. 1960 "Extensionality", *Mind* 69: 55–62, reprinted in L. Linsky, ed., *Reference and Modality*, 1971, 44–51.
Marcus, R.M.B. 1962 "Modalities and Intensional Languages", *Synthese* 14, 2–3: 132–143, repr. in R.B. Marcus 1993.
Marcus, R.M.B. 1988 "Essentialism in Modal Logic", in Marcus 1993, 46–51.
Marcus, R.M.B. 1993 *Modalities*, Oxford, Oxford University Press.
Parsons, T. 2000 *Indeterminate identity, metaphysics and semantics*, Oxford, Oxford University Press.
Peacocke, C. 1976 "What is a logical Constant?", *Journal of philosophy* 73, 9: 221–40.
Pedersen, N.J. (2010) "Stabilizing alethic pluralism", *Philosophical Quarterly* 60: 92–108.
Perry, J., ed., 1977 *Personal Identity*, Berkeley and Los Angeles, University of California Press.
Pettit, P. 1996 "Review of Wright, Truth and Objectivity", *Philosophy and Phenomenological Research* 56, 4: 833–890.
Sainsbury, M. 1996 "Review of Crispin Wright, *Truth and objectivity*", *Philosophy and Phenomenological Research*, 56, 4: 899–904.
Salmon, N. 2002 "Identity Facts", *Philosophical topics*, ed. C. Hill, vol. 30, no.1, repr. in *Metaphysics, Mathematics and meaning*, Oxford, Oxford University Press, 2006, chapter 10.
Unger, P. 1980 "The Problem of the Many" *Midwest Studies in Philosophy* 5: 411–67.
Wiggins, D. 1980 *Sameness and Substance*, Oxford, Blackwell, quoted after second ed. *Sameness and Substance Renewed*, Oxford, Oxford University Press, 2001.
Williamson, T. 1994 *Vagueness*, London, Routledge.
Wright, C. 1992 *Truth and Objectivity*, Oxford, Oxford University Press.
Wright, C. 1996, "Responses to commentators", *Philosophy and Phenomenological Research* 56, 4: 911–91.
Wright C.D. 2010 "Truth, Ramsification, and the pluralist's revenge", *Australasian Journal of Philosophy* 88: 265–283.

Erik J. Olsson

Barcan Marcus on Belief and Rationality

1 Introduction

Ruth Barcan Marcus is famous for her influential work in modal logic, the logic of necessity and possibility. Her perhaps most distinctive contribution to the subject, the so-called Barcan formula, stating that if it is possible that something has a certain property, then there is something that possibly has that property, gave rise to an extensive debate concerning the interpretation of systems of modal logic. While most of Marcus's work on the logic of necessity and possibility dates back some 50 years, she has been writing, more recently, on issues of belief and rationality, rejecting received accounts of propositional attitudes and attempts to systematize epistemic logic. This paper discusses this lesser known part of Marcus's contribution to philosophy and logic, which in a recent publication she describes as still being "in a formative stage" (Marcus, 2005). More narrowly still, I will focus on her reply to Kripke on a puzzle of belief for the theory of direct reference.

2 The theory of direct reference

What is the meaning of a proper name, such as the name Barack Obama?

According to description theories of proper names, each speaker associates with a given name a description that picks out, as it were, the bearer of that name. It is here required that the description uniquely determines the name's referent. Thus, when a speaker uses the name Barack Obama and in doing so succeeds in referring to a particular object or individual x, he manages to do so because he thinks of Barack Obama as the unique person having a certain property, e.g. that of being the current president of the United States, and x in fact does have this property uniquely. By contrast, a direct reference theory of the kind advocated by J. S. Mill (1867) states that the meaning of a proper name is simply its referent or bearer. This view was revived and further developed by

Lund University

Ruth Barcan Marcus in a paper in *Synthese* from 1961, and later adopted by Kripke and others.[1]

The difference between the two theories stands out clearly when one considers how they deal with identity statements of the type "a = b" where a and b are different names. Take for instance the case of "Cicero is identical with Tully". On the direct reference theory, the name Cicero and the name Tully have the same meaning in virtue of referring to the same individual. So the identity statement simply says that that object is identical to itself, a necessary truth. On the description theory, by means of contrast, the statement "Cicero = Tully" is a mere contingent truth. To see this, suppose that, for a given speaker, the name Cicero means "the most famous ancient orator" and that the name Tully means "the person called 'Tully' by the English". The identity statement in question then expresses that the most famous ancient orator is identical to the person called Tully by the English, which is only contingently true. It is contingent because the most famous ancient author could have been called something else by the English.

Both the description and direct reference theory have their well known advantages and disadvantages. The description theory has the advantage of giving an immediate answer to the question of how a name can refer to a given object. The answer is that it does so via the associated definite description. While the direct reference theory does not give an answer to that question by itself, it can be supplemented with a causal theory of reference that explains reference in terms of a causal chain connecting the user of a name with a previous process of baptizing. A problem with description theories is what definite description to associate with a given name. Famous advocates like Frege and Russell allow for different people to associate different descriptions with a given name, something which may appear intolerably subjective because it makes linguistic meaning relative to a speaker. The direct reference theory is clearly less vulnerable to charges of subjectivism. It has been suggested, e.g. by Susan Haack (1978, pp. 64-65), that the two theories are not necessarily rivals but that they may give complementary accounts of the meaning of proper names. Be that as it may. I shall proceed to discuss a puzzle that arises for the direct reference theory in the context of rational belief.

1 See Burgess (1996) for an extensive discussion of the relation between Marcus's and Kripke's work on proper names and related matters.

3 A puzzle about belief

In his paper “A puzzle about belief” from 1979, Saul Kripke raised the following problem for the theory of direct reference. Suppose that all assent is sincere and reflective so that speakers are not conceptually or linguistically confused. Then the following disquotational principle seems highly plausible:

> (DP) If (a) a normal English speaker assents to ‘p’, and (b) ‘p’ is a sentence of English, then he believes that p.

Now a speaker may assent to

> (i) Cicero was bald.

and seemingly coherently assent to

> (ii) Tully was not bald.

And hence also to the conjunction:

> (iii) Cicero was bald and Tully was not bald.

By the disquotational principle, the speaker since he assents to (iii) must be taken to believe that sentence to be true. Yet, on the theory of direct reference, the names Cicero and Tully have the same meaning which means that the conjunction (iii) is a plain contradiction. It says of something that it both is and is not bald. The example suggests that, on the theory of direct reference, a fully rational person may believe a contradiction without having committed any logical blunder or being subject to linguistic or other confusion.

Here is another example, this time of a bilingual nature. From what he has heard or read about the attractiveness of Londres, Pierre a native Frenchman assents to

> (iv) Londres est jolie.

Pierre now emigrates to England where he takes up residence in a neighborhood in London which he finds ugly. There he learns English by exposure, without recourse to translation manuals, dictionaries and the like. He now assents to:

(v) London is not pretty.

Hence, he would also assent to:

(vi) Londres is pretty and London is not pretty.

By a similar (bilingual) disquotational principle, Pierre believes the conjunction (vi) which, by the theory of direct reference, is a plain contradiction saying of London that it is both pretty and not pretty. Pierre exemplifies the predicament of a person who has fallen into contradiction through no fault of his own.

4 Marcus's proposed solution

Kripke views the situation as deeply unsatisfactory. "We may suppose that Pierre ... is a leading philosopher and logician. He would *never* let contradictory beliefs pass ... He lacks information, not logical acumen. He cannot be convicted of inconsistency." (Kripke, 1979, p. 251). Marcus, too, in her 1981 paper, insists that we cannot rest content with this state of affairs. One way of solving the puzzle would be to give up the theory of direct reference in favor of a descriptive theory according to which, as we saw, proper names have a meaning distinct from their reference. On the descriptive theory, the names Londres and London may mean different things for Pierre in virtue of being associated with different definite descriptions. For Pierre, "Londres" could mean "the town I read about in my geography book" and "London" could mean "The town in which I live". If so, the statement "Londres is pretty and London is not pretty" will be false but not contradictory. It will be false in virtue of the contingent fact that the town Pierre read about happens to be the town in which he now lives. That a fully rational person should entertain some false but non-contradictory beliefs is unfortunate but hardly surprising.

Marcus is not inclined to give up the theory of direct reference in response to Kripke's puzzle. Instead, her proposal amounts to arguing that it is the disquotational principle that is ultimately to be blamed for the untoward result. As a preliminary, she rejects the common view that believing is a "propositional attitude", in the sense of an attitude the believer has to a proposition, in favor of the proposal that the object of belief is a *state of affairs*. She goes on to suggest the following condition, which I shall refer to as the *belief-possibility thesis*:

(BP) If X believes that p, then possibly p.

According to this principle, belief is not an entirely internal matter, but what one believes, and can believe, depends on what is, in an external mind-independent sense, possible. As Marcus points out, this part of her proposal consists in "strengthening the connection between belief and 'reality', i.e. possible worlds or structures" (Marcus, 1981, p. 507).[2]

The following modified disquotational principle now suggests itself:

> (DP*) If (a) a normal English speaker assents to 'p', (b) 'p' is a sentence of English, and (c) p *is possible*, then the speaker believes that p.

The effect of this modification is a weakening of the connection between assertion and belief. Assent carries over into belief only if the state of affairs in question is possible.

How does this solve the puzzle about belief? Consider again the case of Pierre, who would assent to "Londres is pretty but London is not", which on the theory of direct reference is a contradiction. On the old disquotational principle, bilingually adjusted, we would be forced to ascribe to Pierre a contradictory belief. This is not so, of course, on the modified principle. Since the sentence in question is contradictory it expresses a state of affairs that is impossible, and so we are not obliged to attribute belief in this case.

The bottom line is that the theory of direct reference can be saved if we are willing to say that belief is not a wholly subjective affair, but something that is governed by Marcus's belief-possibility thesis according to which belief in p requires that p express a state of affairs that is possible. But what independent reason is there for taking this thesis to be true? This is the question to which I now turn.

5 The original motivation for the belief-possibility principle

In her original 1981 paper on Kripke's puzzle, Marcus gives two main reasons why her belief-possibility thesis should be adopted. The first is an argument

2 The reference to "structures" is explained on p. 139 in Marcus's 1990 paper. There she writes "[w]e may think of states of affairs as ordered structures of actual objects which include individuals as well as properties and relations."

from linguistic intuition, the second one from analogy. Let us consider the linguistic argument first. Marcus writes (1981, p. 505):

> Suppose that someone were to claim that he believes Hesperus is not identical with Phospherus or that Tully is not identical with Cicero, or that London is not the same as Londres ... It is my (non post-hoc) intuition that on *discovery* that those identities hold, and consequently that the associated name pairs name the same thing, I would *not* say that I had changed my belief or acquired a new belief to replace the old, but that I was mistaken in claming that I *had* those beliefs to begin with.

But many readers would no doubt find that they themselves lack Marcus's reported intuition. From a pre-systematic standpoint, it seems more natural to say simply that we used to believe, falsely, that Cicero is not identical with Tully but that we have now corrected this belief.

Let us instead turn to the other argument, which makes use of what Marcus thinks of as an "analogy" between knowledge and belief (1981, p. 505), the idea being that there is a close parallel in the following sense:

> Just as a condition for knowing that p is that p obtains, so a condition for believing that p is: if X believes that p, then possible p.

Perhaps we can reformulate this proposal more succinctly in this way:

> (KB) Just as knowing that p requires that p *actually* obtains, so believing that p requires that p *possibly* obtains.

Now this is an interesting thought. What (KB) expresses is a formal relationship between knowing and believing, indeed one of considerable appeal. Still, I believe that Marcus is making too much of her parallel when she refers to it as an "analogy". It is not that the *very same* property holds for both knowing and believing. What is indicated is rather a weaker kind of structural relationship which cannot strictly be subsumed under the more demanding concept of an analogy. It must also be said that the appeal of this structural connection derives from its simplicity and elegance, features that are not obviously indicative of truth. Hence, while Marcus's appeal to (KB), or something similar, does constitute an independent argument for her solution to the belief puzzle, the strength of that argument should not be overestimated.

6 Later motivations for the belief-possibility thesis

In a later paper called "Rationality and Believing the Impossible", which appeared in *The Journal of Philosophy* in 1983, Marcus returns to the task of giving independent motivation for her belief-possibility thesis, her first point being that Berkeley apparently took a similar position in his *Principles of Human Knowledge*. Having argued that existence of matter involves a "contradiction", Berkeley goes on to say (in paragraph 54),

> Strictly speaking, to believe that which involves a contradiction ... is impossible .... In one sense, indeed, men may be said to believe that matter exists; that is, they act *as if* the immediate cause of their sensations, which affects them every moment ... were some senseless unthinking being. But that they ... should form thereof a settled speculative opinion is what I am unable to conceive.

Berkeley concludes:

> This is not the *only* instance wherein men impose upon themselves, by imagining they believe those propositions they have often heard, though at bottom they (those propositions) have no meaning in them.

As Marcus notes, Berkeley seems to be claiming that sentences that describe impossible states of affairs are *meaningless*, whereas Marcus herself is only making what is plausibly seen as a weaker claim, namely, that "where a state of affairs is impossible, there is a sense of 'belief' such that an agent is mistaken in claiming that he is in the belief relation to that state of affairs" (1983, pp. 324-5).

In the same paper, Marcus refers, for indirect support, to an experiment described by Donald Davidson on pp. 235-36 in his 1980 book (quoted from Marcus, 1983, footnote 15 on p. 330):

> After spending several years testing variants of Ramsey's theory [of belief] on human subjects, I tried the following experiment (with Merrill Carlsmith). Subjects made all possible pairwise choices within a small field of alternatives, and in a series of subsequent sessions, were offered the same set of options over and over. The alternatives were complex enough to mask the fact of repetition, so that subjects could not remember their previous choices, and pay-offs were deferred to the end of the experiment so that there was no normal learning or conditioning. The choices for each session and each subject were then examined for inconsistencies – cases where someone had chosen *a* over *b*, *b* over *c*, and *c* over *a*. It was found that as time went on, people became steadily more consistent; intransitivities were gradually eliminated; after six sessions, all subjects were close to being perfectly consistent ... apparently, from the start there were underlying and consistent

> values which were better and better realized in choice. I found it impossible to construct a formal theory that could explain this, and gave up my career as an experimental psychologist.

Marcus remarks that, "[i]n Davidson's experiment it is as if the subjects who assented to sentences that are inconsistent declined to carry them over into belief. Such a subject, if my intuition is shared, would not say, 'I once believed I preferred *a* to *b*, and *b* to *c*, and *c* to *a* but now I don't'. He would disclaim having had such a belief." (1983, footnote 15, p. 330).

There are a number of reasons for being dissatisfied with this interpretation of the experiment. First, while we may agree with Marcus that a subject would not say, "I once believed I preferred *a* to *b*, and *b* to *c*, and *c* to *a* but now I don't", this is so for the trivial reason that a subject would not be in a position to say anything about what she previously preferred, or believed she preferred. Recall that as Davidson describes the experiment, it was set up in such a way that "subjects could not remember their previous choices". Second, even if we assume that the subject did recall their previous preferences, which would require another experimental setup altogether, Marcus's interpretation of the experiment makes use of the very "intuition" that that experiment was intended to support. For these reasons, the belief-possibility thesis is not even "supported indirectly" by Davidson's experiment.

7 A problem for Marcus's thesis

It should be noted that, while Marcus seems to think of her account of the object of belief being a state of affairs rather than something propositional as playing a central role in her proposed solution to Kripke's puzzle, these are seen on closer scrutiny to be two independent views. There are, after all, impossible states of affairs just as there are impossible propositions, and neither Marcus nor anyone else has, to my knowledge, produced any convincing argument to the effect that it should be more tempting to exclude belief in the impossible on one account rather than the other, as it would have been if, for instance, there were impossible propositions but no impossible states of affairs. For this reason, I see no point in upholding the distinction between propositions and states of affairs in the present context, although I believe that this distinction can do useful work in other argumentative settings.[3]

But the real problem that I would like to raise lies elsewhere, for we may legitimately ask whether Marcus's proposal really solves the problem at hand. To be sure, her move does prevent us from ascribing belief in a contradictory sentence to a perfectly rational person. It is still true, though, that our Pierre believes two things that are jointly incompatible, even if he does not believe a contradictory sentence. Pierre assents to each of "Londres is pretty" and "London is not pretty", and so on the original as well as the revised disquotational principle, he comes out believing two sentences that, given the direct reference theory, together form an inconsistent set. That seems almost as bad as believing in a contradictory sentence. For how can a perfectly rational person who is not in any way confused or logically or linguistically misguided believe two things which, for reasons of logic alone, cannot both be true?

In other words: it is still true that the totality of a perfectly rational person's beliefs can be inconsistent, so that there is no possible world where all the person's beliefs are true. My point is that this may seem almost as baffling as the thought that such an agent could believe a contradictory sentence. Marcus's proposal does not address that remaining difficulty.[4] [5]

---

3 See Marcus (1995) for an extended argument to the effect that the distinction is needed to avoid a questionable exclusion of non-linguals from the domain of creatures having beliefs and desires.

4 Brown (1991), p. 358, argues that Marcus's endorsement of the belief-possibility thesis in fact also commits her to what I have called the strong belief-possibility thesis: "Marcus's argument, if successful, would require us to deny not only that one can belief the impossible but, also, contrary to her intentions, that one [can] have contradictory beliefs." See also Altrichter (1985) for a related point.

5 Furthermore, it is still true that a perfectly rational person may fail to believe in the logical consequences of what he believes. A person may assent to "Cicero was bald" without assenting to "Tully was bald", although, on the theory of direct reference, these two sentences mean the same thing and hence should be logically equivalent, and so the one should follow from the other. It is not just the case that logical closure fails when conjoining two beliefs results in contradiction. That logical closure fails in this way is a consequence of Marcus's proposal. But it is also the case that it fails when there is no apparent inconsistency. To make this quite clear: suppose that the person assents to "Cicero was bald" but assents to neither "Tully was bald" nor "Tully was not bald", perhaps because he has no idea who Tully was. In that case, logical closure fails without any inconsistency being inflicted. This problem, too, is not addressed by Marcus's approach, but one may hold it to be less severe than the inconsistency problem, which is why I shall not here pursue this line of thought any further.

8 The strong belief-possibility thesis

Fortunately, there is a quick fix of Marcus's proposal that would block the more general inconsistency problem to which I just drew attention. We recall that Marcus proposes the following belief-possibility thesis:

> (BP) If X believes that p, then the state of affairs that p is possible.

The best motivation for that principle is, again, the following proposal for a structural relationship between knowing and believing:

> (KB) Just as knowing that p requires that p *actually* obtains, believing that p requires that p *possibly* obtains.

But it seems just as plausible to suggest that

> (KB*) Just as there is a possible world (namely the actual world) where everything *known* by a person is true, so there is a possible world where everything *believed* by a person is true.

Hence,

> (SBP) There is a possible world where everything believed by a person is true.

This strong belief-possibility thesis seems no less reasonable than the weaker thesis proposed by Marcus. In both cases, we may draw on an attractive formal relationship between knowing and believing.[6]

---

6 Here is a slightly different argument. The following is true of knowing:

(K) If X knows that p, then there is a possible world w (the actual world) such that p is true in w.

On the basis of this property, we might postulate, with Marcus, that something similar should be true of believing, namely

(BP) If X believes that p, then there is a possible world w such that p is true in w.

The point of course is that the stronger belief-possibility principle blocks the general inconsistency problem. On the stronger principle, for the person to believe separately that Cicero is bald but Tully is not there has to be a possible world where both these sentences are true. But this is prohibited by the theory of direct reference, according to which the names Cicero and Tully have the same meaning in virtue of denoting the same individual in all possible worlds. The same individual cannot be both bald and non-bald.

While the original belief-possibility thesis suggested a natural modification of Kripke's disquotational principle, this is not so for the stronger version. One might think that something like this would work:

> (DP**) If (a) a normal English speaker assents to 'p', (b) 'p' is a sentence of English and (c) p *is possible given the speaker's other beliefs*, then the speaker believes that p.

But suppose again that we come across a speaker who assents separately to "Cicero was bald" and "Tully was not bald" in that order. Using the proposed revised disquotational principle, we would first ascribe to the speaker belief in "Cicero was bald", assuming that sentence to be consistent with his other beliefs, but not ascribe belief in "Tully was not bald". But if we switch the order of the sentences so that "Tully was not bald" is first assented to, we get the opposite result that the speaker believes that Tully was not bald but not that Cicero was bald. This dependence on the order of evaluation is surely hard to swallow.

My own view on the matter is that there is no simple relationship between assent and belief. A person may assent to a given proposition, and yet non-verbally act as if the opposite were true. In many such cases we are inclined to assign greater weight to her non-verbal actions. In her later works, Marcus has

---

But what we have observed, in effect, is that knowing actually satisfies a principle stronger than (K):

(K*) There is a possible world w (the actual world) such that, if X knows that p then p is true in w.

If we follow Marcus in her attempt to strengthen the structural similarity between knowing and believing, we might want to require in addition to (BP) that the following hold:

(SBP) There is a possible world w such that, if X believes that p, then p is true in w.

The latter is of course our stronger belief-possibility thesis.

expressed doubts concerning any disquotational principle that asserts “a privileged and overriding status of certain speech acts as belief indicators in language-using agents” (1983, p. 335). In the 1983 paper she is still apparently undecided concerning the correct notion of rationality. In the paper from 1990, however, she expresses unequivocal support for what she calls a “broader notion of rationality” (p. 138), according to which non-verbal behavior also counts as an indicator or counter-indicator of belief. On this new view, “[i]t is not supposed that the act of sincere assent even where evoked must be an overriding indicator of belief” (1990, p. 141).[7] I believe therefore that she would not consider the mere failure of the strong belief-possibility thesis to provide the basis for a simple principle linking assent to belief to be a convincing argument against that stronger thesis.

9 Consequences for the theory of belief revision

The strong belief-possibility thesis, if accepted, has implications that go beyond the dispute concerning direct reference. The discussion concerning whether or not a fully rational believer can end up believing an impossible proposition is not uniquely tied to identity cases. As Isaac Levi has convincingly argued, observation is plausibly construed as routine expansion whereby a proposition is accepted in a routine fashion (e.g. Levi, 1980). This may, in his view, lead to inconsistency in the set of full beliefs, for what comes to be believed in this fashion, p say, may contradict beliefs that the agent has acquired previously.

7 In her 1990 paper, Marcus advances a dispositional account of belief according to which “x believes that S just in case under certain agent-centered circumstances including x’s desires and needs as well as external circumstances, x is disposed to act as if S, that actual or non-actual state of affairs, obtains” (p. 140). For a discussion, see Engel (1999). This account is used as part of an account of belief that does justice to the plausible pre-systematic intuition that animals and other non-linguals can also have beliefs (and desires). I wonder, though, whether Marcus is not guilty of overkill. She advances an account of belief having essentially two components. Belief is construed as, on the one hand, being a disposition and, on the other hand, as taking something non-linguistic as its object. It seems to me that either component in isolation would actually do the job of securing the possibility of belief and desire in non-linguals. Non-linguals (like lower animals) can obviously be disposed to act, and so they can have beliefs in that sense. If Marcus is correct, non-linguals can also stand in relations to non-linguistic facts, so this condition, too, is satisfied. If so, why invoke *both* a dispositional *and* a non-propositional condition on belief?

So, the agent may come to believe both that p and that non-p. Given logical closure of beliefs, she will end up believing a blatant contradiction.

Now it may be claimed that what is special about the identity cases is that the agent will be unable to detect the impossibility, even if he or she is logically omnipotent. But that this is a genuine difference could be questioned; for it could be argued that the agent, while in the inconsistent state (there is only one inconsistent state assuming beliefs to be closed under logical consequence), is unable to pursue any rational inquiry at all, including inquiry aimed at detecting inconsistency. In a recent debate with Levi in the journal *Synthese*, I have argued that this is indeed a consequence of Levi's theory of routine expansion in connection with his claim that the inconsistent belief state is useless for inquiry and deliberation, that it is "epistemic hell".[8]

This is a troublesome consequence. If inconsistency means hell, how can it ever be legitimate for a rational person to enter that state, and on what basis could consistency be regained? It seems that an inconsistent belief state, if there were such a thing, would be an intellectual point of no return. I suspect that philosophers defending the possibility of having inconsistent beliefs based on our supposed intuitions about the matter underestimate the grave implications of their view (e.g. Altrichter, 1985).

Central in Levi's earlier theory was his claim that it is possible to give principled advice for how a person should extricate herself from inconsistency. His theory of "coerced contraction", as detailed in his 1991 book, was intended to provide such guidance. However, as I pointed out in my *Synthese* paper, his proposal simply won't work: again, once in epistemic hell, there is no rational basis for inquiry, and this goes in particular for inquiry and deliberation into how to escape inconsistency.

In his response to my criticism, Levi devised a new theory of contraction from inconsistency which, unlike his earlier attempt, makes extrication from inconsistency a matter not of deliberation but of routine, the idea being that the agent can precommit herself to a routine for how to handle inconsistency. The routine will automatically, as it were, catapult the agent from the inconsistent state without there being any need for deliberate efforts on his or her part.[9] Yet, there are reasons to be seriously unsatisfied with Levi's new proposal as well, if only because it seems psychologically unrealistic to think that people would be equipped with such elaborate contingency plans for inconsistency-handling. Also, I fail to see how such a routine could be preprogrammed for taking care of

8 See Olsson (2003). The expression "epistemic hell" stems from Peter Gärdenfors (1988).
9 See Levi (2003).

all eventualities, that is to say, all the various ways in which an inconstancy may arise. Considering the grave consequences of ending up in epistemic hell, extensive contingency planning would be imperative. A detailed assessment of Levi's new theory would require a longer discussion, which is best left for another occasion.

This is where Marcus's belief-possibility thesis in its stronger form comes in. Perhaps we should, in Marcus's spirit, think of the inconsistent state not as *epistemic* hell but as something that is not epistemic or doxastic at all. Perhaps, then, an inconsistent state should not be regarded as a state of *belief*. That would open up for a new solution to Levi's long-standing problem of inconsistency, a solution that *prima facie* requires less delicate footwork than Levi's own recent proposal. It would be a radical solution, to be sure, and yet, to paraphrase a famous fictional detective, when all the non-radical possibilities have been examined, the remaining possibility, however radical, has a good chance of being true. At the very least, this approach does seem worthy of serious consideration.[10]

How would the proposal for handling belief-contravening observations work in practice? Suppose the inquirer, while being in belief state K in which non-p is believed, routinely expands by p by relying on a trusted source. What is the new state of belief after the routine has been carried out? On the current proposal, that belief state cannot be the inconsistent state for there is no such state of belief. We may instead ask what *consistent* belief state most plausibly reflects the current commitments of the inquirer. Presumably, that belief state is still K but now augmented with the further belief that the routine in question was invoked giving a certain result, p, that is inconsistent with K. Augmenting K with this information of a report character does not by itself make the new belief state inconsistent. The inquirer may well be in a state in which non-p is believed and also believe that one of his or her previously trusted routines gives a conflicting result.

10 A drawback of this suggestion is that the nice Boolean algebraic structure of potential states of full belief is sacrificed. We cannot anymore think of the space of potential state of full beliefs as featuring two designated states, 1 and 0, and being closed under the two operations of "meet" and "join" (Levi, 1991). While there is still a 1 element corresponding to the belief state in which only tautologies are held true, there is no 0 element corresponding to the contradictory belief state. Since the join of two consistent potential belief states may be inconsistent, closure under it must also be sacrificed, whereas closure under the meet operation can be retained. If potential belief states are represented as sets of sentences of some regimented language, the meet of two such states corresponds to their intersection, and the join of the states to their union.

10 Conclusion

I have argued that Marcus's intriguing proposal for how to solve Kripke's puzzle about belief in the context of a theory of direct reference, while not fully satisfactory as it stands, can be amended in a way that takes care of a remaining problem. The amendment amounts to a strengthening of what I have called her belief-possibility thesis. According to the stronger thesis, a fully rational person's beliefs must be jointly possible in order to count as beliefs at all. Once this has been sorted out, Marcus's general observation still stands: we may continue to endorse the attractively simple theory of direct reference on the condition that we accept a tighter connection between belief and reality. There are, not surprisingly, powerful arguments from the standpoint of ordinary language against such a theory of belief. Altrichter (1985) produces a long list of counter intuitive consequences of Marcus's view, some more serious than others, whereas that view receives a more sympathetic treatment in Brown (1991). I believe that what has been said above adds some weight to the Marcus-friendly side of the dispute. As I have argued, the view that beliefs are by their very nature consistent is in line with a philosophically robust view of inconsistency as a state in which all coherent inquiry and deliberation would break down, so that there would be no rational deliberative escape route. That this is so has, I believe, gone largely unnoticed by Marcus's critics, who have been eager to ascribe inconsistent beliefs in a number of cases, often in an effort to explain action. What they have forgotten is to explain how consistency was, or at least could have been, regained. So, while Marcus's conception of belief is to some extent "revisionary", to use her own characterization in the 1990 paper,[11] it may still be a fruitful explication in the sense of Carnap (1950). Its systematic advantages may in the end outweigh any doubts that arise from what we would say in various cases. A fuller investigation into which view is to be preferred, all things considered, will have to await another occasion.[12]

---

11 Altrichter (1985), as mentioned, emphasizes the divergence between Marcus's account of belief and presystematic intuition on the matter.

12 An earlier version of this paper was read at the 3rd International Lauener Symposium on Analytical Philosophy in Honour of Prof. Ruth Barcan Marcus in Bern 2008. I would like to thank the members of the audience for their comments, including Dagfinn Føllesdal, Timothy Williamson and, above all, Ruth Barcan Marcus herself. I consider myself fortunate to have had the opportunity to meet Ruth, a remarkable philosopher and person, before she passed away.

References

Altrichter, F. (1985), "Belief and possibility", *Journal of Philosophy* 82 (7): 364-381.
Brown, C. (1991), "Believing the impossible", *Synthese* 89 (3): 353-364.
Burgess, J. P. (1996), "Marcus, Kripke, and names", *Philosophical Studies* 84 (1): 1-47.
Carnap, R. (1950), *Logical foundations of probability*, Chicago University Press.
Davidson, D. (1980), *Essays on actions and events*, Oxford University Press.
Engel, P. (1999), "Dispositional belief, assent, and acceptance", *Dialectica* 53 (3-4): 211-226.
Gärdenfors, P. (1988), *Knowledge in flux: modeling the dynamics of epistemic states*, MIT Press.
Haack, S. (1978), *Philosophy of Logics*, Cambridge University Press.
Kripke, S. (1979), "A puzzle about belief", in Margalit, A. (ed.), *Meaning and Use*, Dordrecht, 239-283.
Levi, I. (1991), *The fixation of belief and its undoing*, Cambridge University Press.
Levi, I. (2003), "Contracting from epistemic hell is routine", *Synthese* 135: 141-164.
Levi, I. (2004), *Mild contraction: evaluating loss of information due to loss of belief*, Oxford University Press.
Marcus, R. B. (1961), "Modalities and intensional language", *Synthese* 13: 303-322.
Marcus, R. B. (1981), "A proposed solution to a puzzle about belief", *Midwest Studies in Philosophy*: 501-510.
Marcus, R. B. (1983), "Rationality and believing the impossible", *The Journal of Philosophy 80* (6): 321-338.
Marcus, R. B. (1990), "Some revisionary proposals about belief and believing", *Philosophy and Phenomenological Research* 50, supplement: 133-153.
Marcus, R. B. (1995), "The anti-naturalism of some language centered accounts of belief", *Dialectica* 49 (2-4): 113-129.
Marcus, R. B. (2005), "Ruth Barcan Marcus", in Hendricks, V. F., and Symons, J. (eds.), *Formal Philosophy*, Automatic Press, 131-140.
Olsson, E. J. (2003), "Avoiding epistemic hell: Levi on pragmatism and inconsistency", *Synthese* 135: 119-140.

Joëlle Proust

Ruth Barcan Marcus on Believing Without a Language

Abstract

In her 1990 article entitled "Some Revisionary Proposals about Belief and Believing", Ruth Barcan Marcus reviews the various limitations and puzzles generated by a language-oriented account of belief. The present chapter discusses an extension of Barcan Marcus' revisionary proposal; granting that believers do not need to express their beliefs linguistically, is it justified to consider that a believer may form *reflexive epistemic beliefs* without needing to express them linguistically? Evidence collected in Comparative Psychology suggests that some nonhuman animals can evaluate their own ability to correctly perform a mental task (such as categorizing, or remembering). Granting that this evidence is sound, this raises the question of the representational format in which metacognition has developed in non-humans. The hypothesis developed and discussed here is that metacognition is represented in a non-conceptual, feature-based system, whose function is to evaluate a mental affordance as being incident at a time.

In her 1990 article entitled "Some Revisionary Proposals about Belief and Believing",[1] Ruth Barcan Marcus reviews the various limitations and puzzles generated by a language-oriented account of belief. If "believing that S" is equivalent to "holding a certain sentence true", as Davidson proposes, non-language users cannot have thoughts, a "baffling claim" indeed (134). Another controversial claim of Davidson's is the view that a creature can only have a belief if she can grasp the possibility of being mistaken – only if she possesses the concept of belief. How then, Ruth Barcan Marcus observes, can a preverbal

Institut Jean-Nicod, Department of Cognitive Studies, École Normale Supérieure, Paris

1 See also Barcan Marcus (1981), (1983), (1995).

child be disappointed when she realises that the footsteps she hears are not those of the anticipated person? It seems *prima facie* unjustified to deny a non-verbal organism a disposition to form beliefs, or to feel pains, simply because it cannot express them linguistically. Granting that language is necessary for these mental states to be made reflexively available (we will see that this claim too is controversial), why should it be the case that one can only have beliefs, or pains, if one can be reflexive about the fact that one has them? Another version of the linguistic view on belief is offered by Jerry Fodor, who takes belief to be an attitude towards sentences in the language of thought – "linguistic entities placed squarely in the mind" (137). Fodor, Ramsey and Davidson, as well as any proponent of the linguistic account of beliefs and other attitudes, Barcan Marcus observes, have a hard time dealing with unconscious beliefs and acratic actions; these sit uneasily with the view that beliefs have the form of assented sentences (whether in a public or a private language), for an acratic action is more readily explained by the absence of a conscious formulation of the belief guiding the action. Non-verbal behavior seems often to express our implicit beliefs, and help us discover what we actually believe. Thus giving linguistic assent to the sentence expressing P cannot constitute what it is to believe that P.

In a sequence of papers, Barcan Marcus offers an alternative semantic theory of belief contents, in which

> Believing is understood to be a relation between a subject or agent and a state of affairs (not necessarily actual) but which has actual objects as constituents. (1990, 139)

On this view, the contents of beliefs are constituted by states of affairs understood as "ordered structures of actual objects", including individuals, properties and relations. The dispositional account offered on this basis stresses the connection of belief to behavior: believing is a disposition to *act as if* a state of affairs *obtains*:

> D: X believes that S just in case under certain agent-centered circumstances including x's desires and needs as well as external circumstances, x is disposed to act as if S, – an actual or non-actual state of affairs – obtains. (1990, 140)

In contrast with Russell and Stalnaker, who take contents of beliefs to be truth-evaluable propositions – respectively analysed as structured entities and as the set of worlds in which the belief is true, – Barcan Marcus takes the relevant semantic predicate to be one of a state of affairs "obtaining". The idea is to allow semantic evaluation of belief to proceed even in the absence of sentences as truth bearers. This definition allows us to sidestep the various shortcomings of the linguistic conception of belief. The definition deals with unconscious types of belief, which make it more likely for one to act in a certain way, without needing to make one's reasons explicit. It offers a more natural account of rationality, in which coherence of behavior is what drives the need to have coherent beliefs rather than conversely. An interesting consequence of the definition is that in a case of the "London is pretty" variety, the believer does not need to believe a contradiction. She believes in an impossible state of affairs; more exactly, "she is on D disposed to act as if an impossible state of affairs obtained". (1990, 149) This possibility is quite compatible with the subject being rational and attentive to evidence. Given that a believer is not omniscient, it is quite predictable that she may claim to believe some impossibilities, describable as such in a more encompassing belief system (which may become the believer's own system after the necessary revision is made).

Kripke (1979) in introducing the Pierre puzzle, used the disquotational principle in virtue of which x assenting to a sentence S entails that x believes that S (when conditions of sincerity, competence and reflectiveness obtain). Because Pierre assents to both "Londres est belle" and to "London is not pretty", Pierre is taken to believe both that London is and is not pretty. Barcan Marcus revises this strong version of the disquotational principle. In virtue of D, x can actually be disposed to act on the belief that S even though she sincerely assents to a sentence S' incompatible with S. Assenting to a sentence is neither a necessary condition for believing (for a believer may lack verbal language), nor a sufficient condition (for a believer may not be aware of the beliefs that are actually guiding her behavior, and offer false, although sincere reports). Therefore Pierre may well assent to incompatible sentences, while not having the corresponding beliefs. His assent, in this particular case, "does not carry over into a belief" (1993, 60). He might only claim to have those beliefs, and, if he discovers that they are contradictory, revise his claim to this effect.

As already noticed by Pascal Engel (1998), Barcan Marcus' proposed revision of the definition of belief in turn raises the question of a taxonomy of epistemic states that is implicit in the very distinction between a behavior- and a language-centered view. As Engel convincingly shows, on a line similar to Stalnaker's (1986), acceptances must be recognized as a new variety of propositional attitudes, distinct both from beliefs in Marcus' sense and from assents to sentences. Acceptances include an agentive dimension which is lacking in believing and assenting. Accepting is a mental action, in which a thinker deliberately takes a certain content as a premise in her reasoning or planning. Acceptances are thus context-sensitive and may in various circumstances override a more powerful belief the agent has, but that needs to be prevented from influencing reasoning for various reasons (for example because the goal of reasoning is a *reductio*, or because a piece of information needs to be ignored, for example when reconstructing someone else's thinking). Accepting is a capacity that is closely associated with various reflective practices, involving principles of fairness in interpreting others, or the development of prudential modes of thinking and deliberating.

Extending D to reflexive epistemic beliefs

I will discuss another type of consequence of D by raising the following question: once it is recognized that a believer does not need to express her beliefs linguistically, is it justified to consider that a believer may form *reflexive epistemic beliefs* without needing to express them linguistically? The problem here is admittedly of a different nature. For in the original conception of D, a believer collects information about states of affairs in the world, concerning the properties that accrue to objects, the relations that hold between them, etc. Believing is an information-based disposition to *behave*, i.e. to act physically in a certain context, given one's motivations. The direction in which I am interested in extending D is one in which the disposition to act is mental and self-directed. Why is this way of testing the definition of belief a natural one? In brief, the answer is that evaluating the margin of reliability of one's own mental dispositions is the basic function needed for a cognitive system to flexibly control and monitor uncertainty (whether in belief, reasoning, planning, etc.). Managing this kind of uncertainty leads to the selection of certain courses of action, that is, it influences the agents' dispositions to act in the world. This can be shown by considering the association of belief with action which D emphasizes; this pragmatic conception can be recast as saying that the function of belief is to reduce uncertainty concerning the states of affairs that, from the agents' viewpoint, are relevant to their needs and actions. Uncertainty, however, includes two varieties, as Hume observed. Objective, or factual, uncertainty is the source of uncertainty generated by variations in the external world. It is reduced by collecting evidence on the way the world is. A second source of uncertainty, Hume hypothesized (*Treatise*, I, 4, 1), comes from evaluating one's past ability to reach true judgments: having often been mistaken in drawing conclusions reduces the force of one's belief in a particular judgment, adding its own additional probability of error to objective world variability. Let us use the term "subjective uncertainty" for the additional source of uncertainty generated by variations in a thinker's ability to achieve her cognitive goals (forming true beliefs, perceiving, retrieving facts from memory).

Why does any rational agent need to have some way of distinguishing the two sources of uncertainty? Why cannot she, so to speak, treat them at the same level, as merely unwanted noise, or bad luck, that occasionally causes her to fail at a task? The response is obvious: one should not act in the same way when it turns out that the world is abruptly changing, and when it turns out that one is losing a specific cognitive ability that one was exercising before. In the first case, one can learn new regularities, or evaluate the variations with which one will have to cope. In the second, it would be wrong to revise one's beliefs about the world. What needs revision are the beliefs that one has formed about one's own capacities; the rational actions here consist in requesting help, or using cautionary strategies to monitor one's own abilities.

The contrast that we are making does not need to be understood in internalist terms, as Hume tended to do. No Cartesian introspection needs to be invoked in explaining the source of subjective uncertainty. Both types of uncertainty reflect objective properties and states of affairs, and are frequently hard to disentangle in concrete cases. In a physical action, the goal is to transform the world in a way that is sensitive to one's own desires and beliefs. Success in a physical action can be observed when the action is ended. In a mental action, the goal is a mental property, which the agent wants to acquire in order to be successful in her interactions with the world. The mental property in question, then, is of a normative kind. The subject wants to make a correct decision, whether in perception (consider the thought expressed by the words: "Did I perceive well?"), memory ("Is my memory accurate?"), planning ("Am I ready to perform such and such a complex new action?") or reasoning ("Can I solve this problem?", "Was my reasoning sound, adequate?", etc.). Mental actions in this sense are a frequent component of physical actions; planning a trip to Annapurna is an extreme example of how an agent's capacity to critically evaluate her planning can affect her physical goal (and very survival). Answering these questions requires collecting information of a different kind, based on the dynamics of one's prior abilities, and extracting from it a norm calibrated to the tasks' requirements.[2] Subjective task requirements are states of affairs, relating a kind and level of effort to a benefit. Therefore they can be an object of belief, and even play a prominent role in the critical appraisal of one's beliefs.

Now the similarity with Ruth Barcan Marcus' revisionary argument will hopefully start to emerge. Just as a subject does not need to have the concept of belief to form beliefs, she does not need to have the concept of a mental state to

2 Cf. Proust (2013).

correctly evaluate her epistemic success in a mental task. The reason that can be offered in both cases is similar: *you don't need to have the concept of a mistake in order to revise a false belief; you don't need to have the concept of a mental failure in order to control your attention, or to monitor your basic cognitive (memorial, perceptual) dispositions in a first order task.* In both cases, the crucial element consists in the notion of a state of affairs that is recognized as obtaining or not. In both cases, the actual obtaining of the relevant state of affairs is a precondition of the success of a given action. Although there might be organisms that only have the ability to represent states of affairs that are world-related, it seems that the definition of belief does not exclude those cases where the state of affairs of interest is a disposition of the believer as a mental agent, such as the ability to perform a given first-order task.

Representional form of belief: A problem

The issue raised above, whether a believer may form *reflexive epistemic beliefs* without needing to express them linguistically, has been pressed on philosophers by evidence collected in Comparative Psychology. "Opt out" paradigms have been developed to test animals' ability to monitor and respond adaptively to their own uncertainty.[3] These paradigms offer animals occasional difficult trials; animals' subjective uncertainty is appreciated in their ability to decline to complete them or to seek additional information before responding. Rhesus monkeys, apes, dolphins often produce data patterns in such tasks that are strikingly like those of humans. Furthermore, they don't need to be trained to seek information adaptively in a food-concealment paradigm. Other animals, such as capuchin monkeys, seem to be totally unable to cope with any task of this metacognitive kind.

Now, it is instructive to see that proponents of a sentential view of higher-order mental state representation tend to reject this evidence (Carruthers, 2008, 2011). The line of reasoning is the following. Granting a propositional representation of states of affairs, subjective uncertainty cannot be expressed in thought without forming recursively a first-order representation (e.g. "this is an F") subsumed under a second-order representation (e.g. "my perceiving/judging/believing that this is an F"), itself having the property of being uncertain with

3 Smith et al. (1998), Call & Carpenter (2001), Hampton (2004), Kornell et al. (2007), Beran et al. (2012)

degree *p*, or of being comparatively less reliable than some other representation formed in the past. Developing in full what needs to be thought to self-attribute a degree of confidence for some perceptual judgment thus leads to attributing to the thinker the following minimal conceptual equipment:

1. The capacity to form a first-order representation, whose verbal equivalent is "O is F"
2. The capacity to form the metarepresentation of an epistemic or conative attitude directed at that content, such as, "perceiving (believing etc.) that O is F"
3. The capacity to attribute to the metarepresentation a property that qualifies its relation with the first-order representation : (e.g.:"I perceive that there is a visual display of category A")
4. The capacity to judge that I perceive with uncertainty of degree r that there is a visual display of category A
5. The capacity to attribute the first-order, second-order and third-order representations to myself as one and the same thinker of these representations: PA2 (=judging) PA1(=perceiving, with uncertainty r) [formed by self] (that O is F)

This analysis helps clarify the various "mental" concepts (concepts of mental states) that need to be in a thinker's repertoire to make a fully explicit statement of her uncertainty, and to communicate to others her degree of belief. What makes it deeply unattractive, in the case of animal metacognition, is that it is incompatible with what we know of macaques' (and dolphins') metarepresentational abilities. *According to present evidence, macaques have no mental concepts, do not read minds, and cannot metarepresent that they perceive or that they judge that P.*[4]

Now a conception of belief centered on states of affairs, as D is, may seem to escape the problem of having to attribute a metarepresentational capacity in order to have access to reflexive epistemic properties. A natural suggestion would thus be to merely ignore the requirements of a sentential approach. The question then would be that of the representational structure of belief. A first suggestion, endorsed in Stalnaker's (1987), is that the ways in which propositions are represented "don't matter"; propositions do not need to be assembled out of individuals, concepts, and properties in order to have representational

---

4 In Proust (2013), I offer various arguments to the effect that metacognition does not need to have a metarepresentational structure.

content. Actually, in the pragmatic picture defended by Stalnaker, the *primary* objects of attitudes are "not propositions but the alternative possible outcomes of agents' actions, or more generally alternative possible states of the world". As Stalnaker emphasizes,

> The form in which beliefs and desires are represented is not essential to their content. Two different agents might have the same beliefs even if the forms in which the beliefs are represented are radically different. The conceptual separation between form and content is, I think, the central feature which distinguishes the conception of thought implicit in the pragmatic picture from the one implicit in the linguistic picture. (1987, 23)

This "possible world" picture has its own classical difficulties, associated with the fact that believers may fail to recognize a given state of affairs as being the same in two different contexts, or fail to grasp the consequences of a possible state of the world in a deductively closed way. It is generally considered that although the possible world approach can go some way towards answering these problems (as does Barcan Marcus' distinction between believing and claiming to believe, as we saw above), it may not be equipped to address fine-grained issues that are raised in explaining perspectival facts influencing behavior.

Barcan Marcus' notion of a structured state of affairs may not suffice to articulate belief content for a similar reason. Let us consider again the comparative evidence summarized above. Let us take two agents, a macaque M and a human being H, each representing to him/herself the fact that she does not remember what the color of an icon was. Is there a single state of affairs that is believed by M and H? Definition D takes it that M and H believe that S if they are disposed to act as if S obtains, namely if they decline to respond; they form a similar subject-centered belief (allowing, of course, for their being different subjects). But there are interesting differences in the way they are disposed to act. The states of affairs that are the object of M's epistemic beliefs are not beliefs about the mental in the same sense as in H's case: they do not generalize to other individuals; they do not motivate new procedures to cope with poor memory, etc. In contrast, H's epistemic beliefs are used to compare memory performance over time and across individuals, to take corrective measures to prevent memory loss, etc. Given the conceptual link that the pragmatic view establishes between belief and disposition to act, it seems that a shortcoming of the "structured state of affairs" view of belief is that it fails to account for the differences between M's and H's dispositions to act.

Maybe a response to this worry could be articulated by taking into account "the agent-centered" and "external circumstances" that determine a disposition to act: phylogenetic endowment, learning, (etc.) might explain why a given state of affairs can be used differently by M and H. This is certainly a possible and correct response, but one that is importantly incomplete; it leaves it mysterious how learning or evolution might allow a creature to develop a strikingly different set of inferential dispositions for one and the same type of state of affairs – why should M fail to use S in the way H does, against his own best interests?

Let us summarize the difficulty. With a view on which beliefs are propositions with a quasi-linguistic structure, we cannot account for the evidence of metacognitive beliefs in non-humans. Given a view on which beliefs have no structure, or have a structure inherited from a corresponding state of affairs, we cannot account for the fact that non-humans don't use their metacognitive beliefs as humans do. A way out of this problem that is worth exploring is that there are two varieties of belief having as their contents one and the same state of affairs, differing however in the form they take and, therefore, in the inferential pattern they have.

States of affairs, propositional and non-propositional contents[5]

One way of trying to solve the difficulty summarized above is to hypothesize that metacognition in non-humans is conducted in a representational format that does not license conceptual generalization as it does in humans. Although the same states of affairs form the content of belief in M and H, these beliefs do not influence the inferential system in the same way. Possession of a different way of accessing content and a different way of influencing behavior would justify the claim that there are two forms of epistemic attitudes. Let us see how such a justification might go.

Let us consider first how the issue of having two different formats for expressing the same content can be addressed. Frege and Strawson have emphasized, in their different styles, that the logical structure of predication comes with a metaphysics: the world appears to be composed of independent particulars as bearers of properties and relations, which themselves are dependent universals. A propositional format offers a general framework for referring to

5 This section summarizes a discussion about animal cognition, which appeared in 2009 in Lurz (ed.) and in Proust (2013).

objects, and to truth values, in a unified spatio-temporal system. Strawson therefore calls forms of languages with this structure "particular-based representational systems" (from now on: PBS).[6]

Thanks to propositional representations, humans can recognize the same objects, or agents, including themselves, at different times and locations, and thereby stabilize their knowledge into rich inferential patterns. They can draw on this inferential structure to control their desires, to justify their beliefs, and to plan or explain their actions. Thus propositional thought is an adequate, and probably unique medium for applying concepts to perceptual experience, for combining them into plans and theories, for inferring unobserved properties and events, for articulating the reasons for one's actions, and for expanding one's knowledge.

Rudolf Carnap examined the possibility of having languages with different representational capacities. "Enrichment" is the process through which a thought that is formed in a more basic format is redescribed in the terms of a more sophisticated one, i.e. offering more expressive and conceptual possibilities in terms of descriptive and inferential scope. For enrichment to proceed, there must be a syntactical correlation mapping the representational elements of one structure to the other:[7] the two structures have to be isomorphic under a certain interpretation scheme. Such an interpretation scheme, however, may miss some aspects of the original representational structure, having to do with its relation to context, or to its specific ways of parsing content.

Carnap's notion of enrichment can be generalized to cases in which mental contents don't need to be expressed linguistically. It is quite natural to assume that the analysis of animal metacognition in metarepresentational terms, described above, is a case of enrichment, relative to another, still hypothetical, format. Let us see why this alternative format does not belong to the propositional variety.

The way animals represent states of affairs can be contrasted with that of humans in two respects. At least some animal species may not have any way of re-identifying objects (or themselves as individual beings) as the same over time: their representational system does not respond to the *principle of objectivity*. Let us use the term 'protoconcepts' for the protosymbolic classifiers that

---

6 Basic particulars are reidentifiable, independent entities: material objects or persons. Universals are either sortals (often expressed by common nouns allowing one to count particulars, as in "three apples") or characterizing universals (expressed by verbs and adjectives, which are used on the basis of some prior categorization principle). Strawson, (1959), 168.

7 See Carnap (1937), 224.

these animals use to categorize properties and events, and to possibly infer from them other properties without meeting the objectivity constraint. Protoconcepts, by definition, fail to subsume individual entities because animals cannot re-identify independent objects. If, however, protoconcepts do not apply to individual, numerically distinct property-bearers, they fail to be "strictly determined", as concepts normally are (*vague* attributes excepted). As Frege made clear (following Kant), for any individual, it must be the case that either it falls, or it does not fall, under a given first-order concept. Vague concepts do not have this property; that is why they pose a serious problem for propositional thinking. *Protoconcepts, having no individuals in their scopes, present a property similar to vagueness: they fail to be "well-determined"*. Having no clear-cut boundaries, they possess, rather, similarity-based conditions of application. The protoconcept of [prey], for example, will apply to visual patterns similar to a prototypical prey pattern. This in turn makes it questionable to say that a protoconcept *truly* applies to some particular (currently perceived) pattern. It would be more adequate to say that protoconcepts are more or less efficient classifiers: they have conditions of efficiency, without being truth-evaluable.

A second difference concerns the scope of protoconcepts. Protoconcepts with no objectivity cannot fulfill the generality constraint,[8] i.e. the capacity to generalize predicates across particulars, and reciprocally. If an animal cannot represent negation, quantification, hypothetical reasoning, in association with the generality constraint, its dispositions to act will be substantially reduced even if the same state of affairs is believed to obtain.

These two differences are compatible with the hypothesis that the representational system used in ancestral representational processes is *featural* rather than propositional. Having dealt with this hypothesis in Proust (2009, 2013), I will summarize it here in order to discuss a plausible extension of D to this format. "Placing a feature" has been identified as a basic cognitive competence that can be exercised without concept possession, generality, or objectivity[9]. A feature, as opposed to a property, can be represented as exemplified or "incidental"[10] with no sense of a contrast between a representing subject and a represented object. A minimalist view of features takes them to be close to Gibson's "affordances", i.e. informational patterns with survival significance. Features, qua affordances, belong to an ontology where no subject-world

8 See Strawson (1959), Dummett (1973), and Evans (1982).
9 See Cussins (1992), Campbell (1993), Dummett (1993).
10 Glouberman, (1976).

division is operational.[11] These patterns inform the animal that something valuable or dangerous needs to be acted upon in a certain way (captured, ingested, fled from). A standard example of a feature-placing sentence (FPS), offered by Strawson, is

(1) "there is (little, much) water"

In this type of sentence, a mass-term ("stuff X") is presented as holding at a given time and at a given place – no individual referent can be specified, no 'completeness' (understood as 'saturatedness'), no place-identification is presupposed. Sortals, also called "count nouns", cannot be expressed in this format. You cannot, for example, count "water". What you can do, however, is evaluate the degree to which an affordance is present.

The representation expressed by (1) is not *a truth-evaluable* belief, for it is not structured propositionally: it cannot be true or false. But, as a representation, it still can be misapplied: it has success conditions, depending on whether the corresponding state of affairs obtains. One can suggest, then, the following basic structure for (1):

(2) "There is here and now some (much, little) drinking affordance".

What kind of belief, then, can an animal have in a FPS? It identifies an affordance at a place, categorizes it for its intensity on a gradient scale, and triggers the associated motor programs.

Our present problem is, however, not to know how objectivity and spatial thinking interact, for metacognition has very little to do with spatial information. We can transform an FP system in order to express what is needed to exercise metacognition. What we will call a "feature-based" system (FBS) *evaluates a mental affordance as being incident (at a time);* the state of affairs represented now is a mental affordance, with a given intensity, such as "I don't

---

11 Affordances are relational, rather than being objective or subjective properties. As Gibson observes, "An important fact about the affordances of the environment is that they are in a sense objective, real, and physical, unlike values and meanings, which are often supposed to be subjective, phenomenal and mental. But, actually, an affordance is neither an objective property nor a subjective property; or it is both if you like. An affordance cuts across the dichotomy of subjective-objective and helps us to understand its inadequacy." (Gibson 1979, 129). In contrast with Gibson's antirepresentationalism, however, we consider an affordance to be an *informational* pattern.

remember the color of the icon! I will reject the task!" Expressed in words, an example of a FB representation would be something like

(3) "There is (poor, excellent) A-ing affordance".

Although A-ing actually represents a current mental disposition (to be exercised as part of a task: perceiving, remembering, calculating, planning, etc.), it does not need to be represented "as mental". The animal may simply represent the degree of the mental affordance through a specialized feeling, which motivates the associated disposition to act. Metacognitive features are presumably presented non-conceptually; the human "tip of the tongue" phenomenon provides a good analogy of what these specialized feelings are like, presenting as they do a situated intensive affordance.

On the proposed view, a featural representational system (whether of the FP or the FB variety) includes a notion of normativity, albeit one that involves neither truth, nor success conditions that are taken to apply to propositions.[12] For certainly some dispositions to act are more efficient than others and tend to be selected because they are more efficient. Therefore, sensitivity to norms of adequacy need to emerge in FPS and FBS for these systems to achieve stability.

To be fully convincing, the case for a featural representation of metacognitive states should be grounded in a set of formal rules. Such rules might include some principles of decision relative to the cut-off point where it is reasonable to act on a feature with a given intensity in a given context. They might also include how to handle negation, or its equivalent. It should be intuitively evident that predictively combining features proceeds through integration of intensities and/or differential equations, and allows for very narrow, context-bound inferences, while predictively combining concepts allows generalization of inferences across contexts. Although this point deserves to be discussed at length, I must leave this to another occasion.

One might object that the distinction between a propositional and a feature-based format is only an "ad hoc" hypothesis meant to justify animal metacognition, and to insulate it from a metarepresentational capacity. There are various arguments, however, that independently speak in favor of this distinction. One argument is related to the need to understand how flexible use of information emerges in phylogeny. Even those non-human animals that, like spiders, have no sense of objectivity, and, therefore, no ability to represent an independent

12 Success conditions can be used in a semantic theory. See Stalnaker (1987) and Bermudez (2003). But success semantics for a non propositional format has not been offered yet.

world, can recognize when a state of affairs obtains, and act flexibly on this basis.[13] Therefore they have to be granted some form of representational structure, allowing them to store and retrieve relevant information. Non-human as well as many human forms of control largely rely on perceptual forms of feature-placing.

A second reason for exploring alternative types of representation is that we need to understand how propositional content has evolved. It makes little evolutionary sense to say that propositional thought appeared with the emergence of linguistic abilities, for linguistic abilities themselves require flexible controls to be exercised. It seems difficult to assume that propositionally structured beliefs directly appeared in those animals, – the vertebrates, some cephalopods – that represent the world objectively. Some story must be told about how objectivity could have come about over evolutionary time.

A third argument in favor of recognizing feature-based representational systems is that this assumption puts the issue of non-conceptual content in a new light. Non-conceptual contents might have been the first way in which states of affairs were represented, and have been subsequently enriched into propositional thought.[14]

Conclusion

The present proposal offers an extension of D that is complementary to Pascal Engel's suggestion for including acceptance as an additional form of belief. While Engel considers cases that are downstream from "simple belief", this proposal works upstream. It tries to uncover a primary type of belief that is already implicit in D. In the proposed reading, a definition of belief along the lines suggested by Ruth Barcan Marcus allows us to discuss the similarities and differences between two ways in which one can act as if S obtained. Non-propositional belief would be present when S is represented in a protoconceptual featural representation system; propositional belief would be

---

13 One might be tempted to object that animals' behaviors, such as the spiders', don't use representations worth of the name, as their decisions to act are strictly based on conditioning mechanisms. This argument has been shown to be wrong since the late 60s: as contemporary learning theorists have shown, associative learning indeed depends on the information that is represented by an animal, it does not bypass it.

14 See Proust (2013, ch.14).

present either when the featural representation is redescribed through conceptual enrichment, or when the representation is directly built as a proposition.

The case for this distinction is based on recent evidence for metacognition in nonhuman animals. Non-conceptual cues might allow an animal to represent a given physical affordance or mental disposition, without needing to represent it as physical or mental. The interest of the distinction, however, goes beyond the clarification of the particular case of epistemic reflexivity. The introduction of this primary form of belief raises two more general questions. First, what is the role of propositional form in determining the functional role of belief? Is it true that, contrary to one type of pragmatic understanding of D, form does matter to belief? In the present proposal, different forms might lead to different types of belief, in the sense that the dispositions to act are in part structured by formal properties of the representational system. Second, assuming that there are several kinds of attitudes, is language a precondition for entertaining specific kinds of attitudes? Could it be the case, as Engel (1999) suggests, that believing does not require a language, while assenting to a proposition, or accepting it pragmatically, does? The present proposal suggests that certain forms of acceptance, such as deciding to act on the assumption that a slightly uncertain belief – or memory – is reliable, require representation of a feature, rather than assent to a sentence. These are some of the fascinating questions that are raised in the wake of Ruth Barcan Marcus' discussion of belief.

Acknowledgement

I owe the basic idea of taking metacognition to have a non-propositional content to my co-worker and IP leader Hannes Leitgeb (University of Bristol) in the ESF project referred to below. His observation that the same formal limitations apply to certain forms of belief revision and to metacognition led me to develop the present view on metacognition as feature-based. I thank Dick Carter for his help in the linguistic revision of the present article, and for his comments. I also thank Jérôme Dokic, Pascal Engel, Claudine Tiercelin, Tim Williamson and the participants of the Lauener Prize Workshop and of the Metacognition seminar for helpful discussions. This work, as part of the European Science Foundation EUROCORES Programme CNCC, was supported by funds from CNRS and the EC Sixth Framework Programme under Contract no. ERAS-CT-2003-980409.

References

Barcan Marcus, R. 1981. A Proposed Solution to a Puzzle about Belief, in P. French et al. (eds.), *Midwest Studies in Philosophy: Foundation of Analytic Philosophy*, 6, 501–10.
Barcan Marcus, R. 1983. Rationality and Deciding the Impossible. *The Journal of Philosophy*, 81, 321–339.
Barcan Marcus, R. 1990. Some Revisionary Proposals about Belief, in *Philosophy and Phenomenological Research*, 50, 133–153. Reprinted in *Modalities*, Oxford: Oxford University Press, 1993.
Barcan Marcus, R. 1995. The Anti-naturalism of some Language-Centered Accounts of Belief. *Dialectica*, 49, 2–4, 113–129.
Beran, M., Brandl, J., Perner, J. & Proust, J. (2012). *The Foundations of Metacognition*. Oxford: Oxford University Press.
Bermúdez, J. L. 2003. *Thinking Without Words*. New York: Oxford University Press.
Call, J., & Carpenter, M. (2001). Do apes and children know what they have seen? *Animal Cognition*, 4, 207–220.
Campbell, J. 1993. The body Image and Self-Consciousness, in J. Bermudez, T. Marcel & N. Eilan (eds.), *The Body and the Self*: Oxford, Oxford University Press, 28–42.
Carnap, R. 1937. *The Logical Syntax of Language*. London: Routledge & Kegan Paul. Carruthers, P. 2008. Meta-cognition in Animals: a Skeptical Look. *Mind and Language*, 23, 1, 58–89.
Carruthers, P. 2011. *The Opacity of Mind*. Oxford: Oxford University Press.
Cussins, A. 1992. Content, Embodiment and Objectivity: The Theory Of Cognitive Trails. *Mind*, 101, 651–688.
Davidson, D. 1980. *Essays on Actions and Events*, Oxford: Oxford University Press.
Dummett, M. 1973. *Frege. Philosophy of Language*. London: Duckworth.
Dummett, M. 1993. *The origins of Analytical Philosophy*. London: Duckworth.
Evans, G. 1982. *The Varieties of Reference*. Oxford: Oxford University Press.
Frege, G. [1892] 1951. On Concept and object, translated by P. T. Geach, Max Black. *Mind*, Vol. 60, No. 238 (Apr. 1951), 168–180.
Gibson, James J. 1979. *The Ecological Approach to Visual Perception*. Boston: Houghton Mifflin.
Glouberman, M. 1976. Prime Matter, Predication, and the Semantics of Feature-Placing, in A. Kasher (ed.), *Language in Focus*. Reidel, 75–104.
Hampton, R. R., Zivin, A., Murray, E. (2004). Rhesus monkeys (*Macaca mulatta*] discriminate between knowing and not knowing and collect information as needed before acting. *Animal Cognition*, 7, 239–246.
Hume, D. [1739–1740]. 1978. *A Treatise of Human Nature*. Oxford: Oxford University Press.
Kornell, N., Son, L., & Terrace, H. 2007. Transfer of metacognitive skills and hint seeking in monkeys. *Psychological Science*, 18, 64–71.
Kripke, S. 1979. A Puzzle about Belief, in A. Margalit (ed.), *Meaning and Use*, Dordrecht: Reidel, 239–283.
Proust, J. 2008. Epistemic agency and metacognition: an externalist view. *Proceedings of the Aristotelian Society*. Meeting of June 2, 2008, vol. CVIII, 3, 241–268.
Proust, J. 2009. The Representational Basis of Brute Metacognition: A Proposal, in R. Lurz (ed.), *Philosophy of Animal Minds: New Essays on Animal Thought and Consciousness*, Cam-bridge: Cambridge University Press, 165–183.

Proust, J. 2013. *The Philosophy of Metacognition: Mental Agency and Self-Awareness*. Oxford: Oxford University Press.
Smith, J. D., Shields, W. E., Allendoerfer, K. R., and Washburn, D. A. 1998. Memory monitoring by animals and humans. *Journal of Experimental Psychology: General*, 127, 227–250.
Stalnaker, R. 1987. *Inquiry*. Cambridge: MIT Press.
Strawson, P.F. 1959. *Individuals*. London: Methuen.

Edgar Morscher

Moral Dilemmas: From a Logical and from a Moral Point of View

Abstract

Prof. Barcan Marcus is renowned for her work in logic, but her contribution to philosophy is by no means restricted to this area. Her 1980 paper on moral dilemmas became a kind of cult paper in the area of moral philosophy, where there is simply no getting round it. On Prof. Barcan Marcus's view moral dilemmas belong inevitably to our moral reality. This moral fact must be taken into account in shaping deontic logic in a way suitable to the purposes of morality. On the other hand, a moral lesson can be drawn from logic, as Prof. Barcan Marcus, whose concern is particularly for young people, emphasizes: Try to lead your life in such a way as to avoid moral dilemmas as much as possible.

Owing to her pioneering work in quantified modal logic, Professor Ruth Barcan Marcus soon made a name among philosophers and in particular among logicians. Quantified modal logic has important applications in different philosophical areas, of which moral philosophy is not the least. Nevertheless, it came as a surprise to many members of the philosophical community when in 1980 Prof. Barcan Marcus published her paper "Moral Dilemmas and Consistency" (from now on quoted as 'A', reprint with 'B'), which soon turned out to be a trailblazing contribution to moral philosophy. There is no serious work on moral dilemmas in our day which does not – explicitly or implicitly – make reference to this paper and its 1996 sequel, entitled "More about Moral Dilemmas" (henceforth quoted as 'C'). In the *first part* of my paper I will summarize the main results of Prof. Barcan Marcus's papers on moral dilemmas, and in the *second part* I will add some comments and questions.

Universität Salzburg

Part I: What is it for a moral code to be consistent, and what is a moral dilemma?

Summarizing a brilliant paper is always a thankless task since the summary will never compare to what is to be summarized. And Barcan Marcus's paper on moral dilemmas is *indeed* brilliant. As with her work altogether, one is tempted to say, modifying a Hitchcock dictum on philosophy: small letters – few words – enormous impact. So let me try to summarize it to the best of my ability.

Traditionally, moral dilemmas were taken to be logical failures in a moral code which are to be erased and eliminated. Barcan Marcus rejects this common treatment of moral dilemmas and conflicts: for her moral dilemmas are inevitable and unavoidable constituents of our moral reality. In order to realize this, we must conceptually separate moral dilemmas from normative inconsistencies, and conflicting norms and duties from those which are logically incompatible. A moral dilemma is not or at least need not be an inconsistency of moral principles, nor need it arise from an inconsistency in a moral code. This insight requires us to make clear what it is for a moral code or a set of moral principles to be inconsistent. In the common treatment of moral dilemmas, which views moral dilemmas as evidence for inconsistencies, the following concept of consistency is usually presupposed:

(1) A set of moral principles is $consistent_1$ iff it applies without conflict to all possible cases (A 123); that is, expressed in terms of possible worlds semantics:
A set *S* of moral principles is $consistent_1$ iff in every possible world all obligations of *S* can be fulfilled.

As long as consistency is understood in this way, a moral dilemma is always a sign of inconsistency and must therefore disappear within a standard logical frame. Thus, Barcan Marcus proposes another concept of consistency for a moral language:

(2) A set of moral principles is $consistent_2$ iff there is a possible world in which they are all obeyable in all circumstances in *that* world (A 128), i.e., iff there are possible circumstances in which no conflict will emerge (A 129); put in possible worlds terminology:
A set *S* of moral principles is $consistent_2$ iff there is a possible world in which all obligations of *S* can be fulfilled.

By negating the definiens of (1) and (2), respectively, we get the corresponding concepts of inconsistency of a set of moral principles, which we will refer to as (1*) and (2*), respectively:

(1*) A set of moral principles is inconsistent$_1$ iff it does not apply without conflict to all possible cases, i.e.:
A set *S* of moral principles is inconsistent$_1$ iff there is a possible world in which not all obligations of *S* can be fulfilled.

(2*) A set of moral principles is inconsistent$_2$ iff there is no possible world in which they are all obeyable in all circumstances in *that* world, i.e., iff there are no possible circumstances in which no conflict will emerge, i.e.:
A set *S* of moral principles is inconsistent$_2$ iff there is no possible world in which all obligations of *S* can be fulfilled.

A moral code, which is consistent$_2$, need not be consistent$_1$; but if a moral code is inconsistent$_2$, it must also be inconsistent$_1$. Weakening the concept of consistency of a moral code (like moving from (1) to (2)) makes it easier for a moral code to remain consistent in spite of the reality of moral dilemmas.

Weakening the concept of a moral dilemma is another move in the same direction. The common treatment prior to Barcan Marcus takes moral dilemmas as situations in which we have an obligation to do *A* and an obligation to refrain from doing *A*, i.e., a prescription and a proscription of *A*:

(3) $O(A) \wedge O(\neg A)$

Given the well-known deontic axiom D, i.e., the axiom $O(A) \rightarrow \neg O(\neg A)$, or, equivalently, $\neg(O(A) \wedge O(\neg A))$, a situation such as (3) automatically results in an inconsistency not only of type (1*) but even of type (2*).

We have first to realize that moral dilemmas do normally not appear in form of (3), but rather in the following form:

(4) $O(A) \wedge O(B) \wedge \neg\Diamond(A \wedge B)$

A situation as described by (4) signifies an inconsistency of type (1*). It alone, however, is no evidence for an inconsistency of type (2*). In order to show this, let us explain what it means that there exists a moral dilemma as represented by (4) in a possible (or also in our actual) world *w*. (4) is expressed in a multimodal language *L* which contains a deontic modality ('*O*') as well as an alethic modality ('$\Diamond$'). A model of a language of this kind is a quadruple $\langle W, R_d, R_a, V\rangle$, containing a non-empty set W of possible worlds, a serial relation R_d on W that serves as deontic accessibility relation, a relation R_a on W that serves as alethic accessibility relation and an assignment function V from the set of formulas of L to the set $\{1, 0\}$ of truth values. That there exists a moral dilemma as represented by (4) in a world w of W can be expressed in the semantic metalanguage of L as follows:

$$V(O(A) \wedge O(B) \wedge \neg\Diamond(A \wedge B), w) = 1,$$

and from this we get:

$$V(O(A), w) = 1, \text{ and } V(O(B), w) = 1, \text{ and } V(\neg\Diamond(A \wedge B), w) = 1.$$

The last component of this conjunction – i.e., $V(\neg\Diamond(A \wedge B), w) = 1$ – is the most important part with respect to the question of consistency; and $V(\neg\Diamond(A \wedge B), w) = 1$ comes true just in case the following holds:

(5) For every $w' \in W$: if wR_aw', then $V(A \wedge B, w') = 0$.

From (5), however, it does not follow that the set $\{O(A), O(B)\}$ is inconsistent$_2$. Quite on the contrary, (5) – i.e. the semantic representation of (4) – is compatible with the set $\{O(A), O(B)\}$ being consistent$_2$, viz. with

(6) There is at least one $w' \in W$ such that $V(A \wedge B, w') = 1$,

whereas (as soon as there is at least one $w' \in W$ with wR_aw') (5) would not be compatible with $\{O(A), O(B)\}$ being consistent$_1$, viz. with

(7) For every $w' \in W$: if wR_aw', then $V(A \wedge B, w') = 1$.

Such analyses of the concept of a moral dilemma and of the concept of consistency (or inconsistency, respectively) account for the insight that a moral dilemma does not rest or at least need not rest on a logical defect of the underlying moral code but *is* or at least *can be* of a contingent origin. (Further fruitful distinctions among moral dilemmas which we will ignore in the present context can be found, e.g., in Sinnott-Armstrong (1996) and in Brink (1996); cf. also Morscher (2002).)

Even if we use a weaker concept of consistency and a weaker concept of moral dilemmas than usual we are not safe yet. Still the reality or mere possibility of a moral dilemma (in the weaker sense of (4)) may result in an inconsistency of type (2*). The question is how to prevent a moral code, which is consistent$_2$ from ruling out the possibility of a moral dilemma in the weaker sense of (4). How can we reconcile moral codes, which are consistent$_2$ with the possibility of moral dilemmas in the sense of (4)?

From her view of moral dilemmas and the consistency of a moral code, Prof. Barcan Marcus draws certain conclusions about how to shape a system of deontic logic that is appropriate to moral facts. These conclusions will be considered in Part II of this paper. Besides this *logical* moral, however, Ruth Barcan Marcus draws also a profound "*moral* moral" – so to speak – from her analysis. She puts it in a second-order principle: Strive to arrange your life and encourage social arrangements that would as far as possible prevent future conflicts from arising (A 133, B 138, C 33). Here we have – so to speak – a deontic ought-implies-can principle, in contrast to the common logical ought-implies-can principle OIC (cf. Part II). We can express this deontic principle formally as follows:

(8) $O(O(A) \rightarrow \Diamond(A))$

This second-order rule has considerable moral impact – with respect to personal morality as well as social, in particular political morality. It cannot be appreciated in its full extent, however, as long as we fail to take moral dilemmas seriously, i.e., as inevitable part of our social reality. It is this very point where her concern about our society and in particular about today's youth turns the logician Barcan Marcus into an educator and moral teacher. It is here that not only morality has a lesson for logic, but also logic has a lesson for morality.

Part II: How to reconcile deontic logic with moral dilemmas

How must a system of deontic logic be shaped in order to meet the criteria of adequacy implied in Ruth Barcan Marcus's analysis of moral dilemmas? Prof. Barcan Marcus herself gives several hints how to do it. In what follows I will underline some of her proposals, comment on some and raise a few questions.

1. Standard systems of deontic logic are obviously not adequate for Barcan Marcus's purpose. Such a standard system of deontic logic, in its propositional version, is based on propositional logic and two specific deontic axioms and a specific deontic rule:

Axiom K: $O(A \rightarrow B) \rightarrow (O(A) \rightarrow O(B))$
Axiom D: $O(\neg A) \rightarrow \neg O(A)$, or, equivalently: $\neg(O(A) \wedge O(\neg A))$

Rule N of Necessitation (for 'O'): From A to infer $O(A)$.

From axiom K and rule N we get immediately a derived rule, which was already a principle of the first system of deontic logic, developed by Ernst Mally; in its semantic version it is often called a deontic principle of logical consequence and stated as follows (whereby '$\vDash$' stands for logical truth):

DPLC: If $\vDash A \rightarrow B$, then $\vDash O(A) \rightarrow O(B)$.

Barcan Marcus rejects explicitly axiom D and DPLC, i.e., the deontic principle of logical consequence (C 29 f.). Another target of her critique is the agglomeration principle (or – as Ruth Barcan Marcus calls it – the factoring principle) which is derivable from axiom K and rule N and is therefore a theorem of standard systems:

AP: $(O(A) \wedge O(B)) \rightarrow O(A \wedge B)$

The only principle she seems prepared to accept is the minimal axiom

MA: $\neg O(A \wedge \neg A)$

A minimal deontic system based on MA (and including also DPLC) was proposed by Brian Chellas (Chellas (1980), pp. 202 and 272 ff.), who placed particular emphasis on the fact that D and AP are not theorems of his minimal deontic logic (Chellas (1980), p. 273).

Barcan Marcus comments on such a minimal system: "In such a system, dilemmas are not ruled out as they are by 1", i.e., axiom D (C 30). Axiom D alone, however, merely rules out moral dilemmas as described by (3), i.e. '$O(A) \wedge O(\neg A)$', but it does not, of course, automatically also rule out moral dilemmas of type (4), i.e. '$O(A) \wedge O(B) \wedge \neg\Diamond(A \wedge B)$'.

2. Whether moral dilemmas of type (4) are or are not ruled out by a logical system does not depend on its purely deontic principles (axioms and rules) alone but also on its treatment of the modal operator '$\Diamond$'. If we would, e.g., interpret '$\Diamond$' as possibility in the strict logical sense and choose our axioms in accordance with this interpretation, a situation as described by (4) would be excluded by logical reasons alone: '$\Diamond(A)$' would in this case be true just in case there is at least one possible world in which A is true; '$\neg\Diamond(A \wedge B)$' would therefore be true iff there is *no* possible world whatsoever in which '$A \wedge B$' is true. The situation described by (4), however, requires that '$O(A)$' is true and '$O(B)$' is true, and therefore – given the agglomeration principle AP – '$O(A \wedge B)$' must be true too; and if '$O(A \wedge B)$' is true, there must be – due to R_d's being serial – *at least one* possible world in which '$A \wedge B$' is true.

Moral dilemmas of type (4) would also be ruled out by our logic if we were to combine the agglomeration principle AP with an ought-implies-can principle of the following kind:

OIC: $O(A) \rightarrow \Diamond(A)$

In this case '$O(A \wedge B)$' would be derivable from '$O(A) \wedge O(B)$' by AP, and '$\Diamond(A \wedge B)$' from '$O(A \wedge B)$' by OIC, in contradiction to '$\neg\Diamond(A \wedge B)$' which is a conjunctive component of (4).

3. In dealing with the question of whether a moral dilemma of type (4) is or is not ruled out by logic we must therefore not restrict our examination to purely deontic systems but must take into account also the non-deontic components of multimodal systems. This is in clear agreement with Ruth Barcan Marcus's view: "... a satisfactory system [of deontic logic] will be very complicated indeed. It will require some very non-standard assumptions to cope with the difficulties of closure. An adequate semantical base for a theory of obligation will require a semantical theory in which sentences designate possible states of affairs. It will also need to include modal operators (logical and metaphysical), temporal operators, and operators for physical modalities, for the latter is required for a proper treatment of the more general ought implies can" (C 31).

Stig Kanger had already developed multimodal systems of this type, including in addition also epistemic operators and operators which are called nowadays '*stit*-operators' (Kanger (2001), esp. pp. 120–185; cf. Morscher (2012), pp. 185–202). He had started his investigations into legal logic and his logical analysis of different types of human rights by using a causing-operator. Since the parties involved in rights can also be legal entities such as states, organizations, companies, institutions etc. which cannot be causes in the literal sense of the word, he very soon switched from 'x causes the fact that p' to the formulation 'x sees to it that p', i.e., to a *stit*-operator, for which we will use here the letter 'S' (whereby '$xS(A)$' can be read as 'x sees to it that A'). Whereas the *stit*-operator differs from the causing-operator with respect to its interpretation, the two operators basically coincide with respect to their logic. Both of the two operators are governed by an axiom of type T, namely

Axiom ST (axiom T for 'S'): $xS(A) \rightarrow A$

For operators such as the *stit*-operator, there is neither a rule of necessitation nor a principle of logical consequence available, since principles like the following ones are obviously *invalid*:

SN: If $\models A$, then $\models xS(A)$.
SPLC: If $\models A \rightarrow B$, then $\models xS(A) \rightarrow xS(B)$.

The problems which seem to endanger the corresponding principles for the O-operator (viz. rule N and principle DPLC), thereby putting a strain on the O-operator, are shifted by Kanger into the logic of the *stit*-operator, thereby saving all that is dear to the logician for the O-operator. Of course, Stig Kanger is not the only author to have made fruitful use of a *stit*-operator in his work in moral and legal philosophy, but he certainly was one of the first to do so. Nevertheless he is only seldom given the credit he deserves for what he did. No reference to him or his works appears often even in the writings of authors who pay special attention to *stit*-operators, such as, e.g., Belnap & Bartha (1995). That is the reason why I am placing here particular emphasis on Stig Kanger's use of the *stit*-operator.

4. Ruth Barcan Marcus's treatment of moral dilemmas provides a definite vote in favour of alternative systems of deontic logic, deviating from the main route of standard deontic systems. In this context it is even more important, on my view, to distinguish different types of normative 'oughts' and to be aware that different types of normative 'oughts' require different sets of logical principles. The main failure of deontic logic is in my view the illusion shared by many of its representatives that one and the same set of logical principles fits all different kinds of normative 'oughts'.

Of course, moral philosophers and deontic logicians have always made an important distinction between ought-to-be and ought-to-do. But the distinction has too seldom found appropriate representation in the formal apparatus of a logical system. By merely using a *stit*-operator we can express, e.g., a difference between two types of moral dilemmas mentioned by Barcan Marcus, i.e., between the moral dilemma for an individual (such as Plato's case, in which the return of weapons has been promised to one who, intent on mayhem, comes to claim them) or a dilemma like the one in the case of Antigone and Creon:

(4a) $O(xSA) \wedge O(xSB) \wedge \neg\Diamond(xS(A \wedge B))$

(4b) $O(xSA) \wedge O(ySB) \wedge \neg\Diamond(xS(\mathrm{A}) \wedge yS(B))$

But not only the expressibility of our deontic language is enlarged by such additions, also some obstacles like the paradox of the good Samaritan will disappear: Even if we accept that 'the robbed person is helped by the good Samaritan' entails 'someone has been robbed', '*x* sees to it that the robbed person is helped' does in no way entail '*x* sees to it that someone has been robbed'. (Incidentally, in my view the so-called paradox of the good Samaritan is not a real paradox at all but only seems to be one, for the following reason: 'someone has been robbed' is entailed by 'the robbed person is helped by the good Samaritan' merely thanks to a Russellian logic of the definite description 'the robbed person', according to which this sentence is analyzed as 'there is exactly one person who is robbed and this very person is helped by the good Samaritan'. Given this reading, we would never maintain that it ought to be the case that the robbed person is helped by the good Samaritan, since this would include maintaining that it ought to be the case that at least one person is robbed. Cf. Morscher (2012), pp. 172–177.)

There is also an important difference within the scope of 'ought-to-do'. There is the biblical saying that, when asked to go one mile, one ought to go the second mile (Matthew 5, 41). In medieval Latin, actions which we *ought* to perform in this sense, were called *opera supererogationis*, from which Chisholm derived the term 'supererogation' (Chisholm (1963); this Chisholmian term is still in use in contemporary moral philosophy). The 'ought' as used in this supererogatory sense has obviously a logic different from that of an 'ought' of strict obligation (cf. Wessels (2002)). And incidentally, Sartre's much-cited example of a moral dilemma (where the choice to be made by the agent is between not abandoning a wholly dependent mother and becoming a freedom fighter) concerns a conflict between oughts, which seem to express acts of supererogation rather than strict obligations.

The upshot of this is that different oughts involved in moral dilemmas require different sets of logical principles and therefore different systems of deontic logic. One cannot just lump them all together.

Despite the fact that Prof. Barcan Marcus does not stress this point in her papers on moral dilemmas, I assume from what she wrote in her earlier article on "Iterated Deontic Modalities" (Barcan Marcus (1966)) that she can subscribe to it.

5. A painstaking investigation brings to light that even *stit*-operators will not be enough. For an evaluation of the agent it is not primarily important what he or she sees to bring about but rather what he or she *intends* to see to bring about. What we need therefore is an *istit*-operator rather than a mere *stit*-operator: x intends to see to it that p, or, to be more specific: by performing act a of type T, x intends to see to it that p. Thereby the basic functor of our deontic logic turns into an inhomogeneous four-place operator, including two individual variables (for individual acts and for agents), a predicate variable (for the type of an act) and a propositional variable.

6. A final point: In all these reflections so far I have assumed that our different logics of different types of 'ought' are "propositional" insofar as the ought is always to be applied to something – however complex it may be – "propositional". As a non-cognitivist myself, I have always denied "normal" truth-values to ought-sentences, but allowed a rational evaluation of ought-sentences in terms of "valid" and "invalid", "1" and "0" etc. But now I think that an uncompromising non-cognitivist could – and maybe also should – give up this presupposition and talk about moral decisions no longer in terms of (propositional) *credibility* but rather in terms of (non-propositional) *acceptability*. All the irregularities appearing in normative discourse again and again (like so-called deontic paradoxes, failure of monotonicity etc.) seem to get a more appropriate explication within such an approach. That, however, is another story.

Special thanks to Peter Simons for helpful improvements.

Appendix 1: Ruth Barcan Marcus on Moral Dilemmas and Consistency

[Reprint of Morscher (1982), pp. 94–97; the notation of logical symbols has been matched with that used in the preceding paper]

In a very illuminating paper, which throws new light on many important questions of moral philosophy, Ruth Barcan Marcus deals with the problem of "Moral Dilemmas and Consistency". This note does not attack her views, to most of which I subscribe. It is rather a short comment on what I take to be the most interesting aspect of Prof. Barcan Marcus's paper.

Prof. Barcan Marcus rejects the common treatment of moral dilemmas or conflicts. This treatment takes their existence to reflect some kind of inconsistency in the set of moral principles or rules, e.g., in the moral code. In viewing moral dilemmas as evidence for inconsistencies, one is usually presupposing the following concept of consistency:

(1) A moral code, i.e., a set of moral rules or principles, is consistent$_1$ if it applies without conflict to all possible cases (p. 123).

Dilemmas, then, are usually resolved by tinkering with the set of moral principles or rules, like hedging them with exception clauses, establishing a rank ordering, etc. Through such a procedure, the original dilemma disappears: it is shown to have a solution without residue and so to have been merely apparent rather than real.

According to Prof. Barcan Marcus, it is much more appropriate to the moral facts if we take moral dilemmas to be real and do not try to spirit them away. Prof. Barcan Marcus proposes a definition of consistency for moral rules different from (1):

(2) A set of moral rules is consistent$_2$ iff there is a possible world in which they are all obeyable in all circumstances in *that* world (p. 128), i.e., iff there are possible circumstances in which no conflict will emerge (p. 129).

What I want to emphasize in this note is that the definition of consistency proposed by Prof. Barcan Marcus in her paper is one which is "normal" or "standard" for non-normative languages; its use for normative languages is novel only in comparison with other proposals for these, which usually deviate from this "normal" definition. Such deviation from the "normal" definition of consistency mostly results from assuming specific deontic laws such as $O(\neg A) \rightarrow \neg O(A)$, or $(O(A) \wedge O(B)) \rightarrow O(A \wedge B)$, or $O(A) \rightarrow \Diamond(A)$, which are far from being as harmless as they might look at first glance. Prof. Barcan Marcus has purified the concept of consistency of such "impurities" and reduced the concept to its real and "normal" kernel.

Let us consider a very simple set $K = \{O(A), O(B)\}$ consisting of the two moral principles $O(A)$ and $O(B)$ (where 'O' stands for 'it is obligatory that' or 'it ought to be the case that'). For the sake of simplicity I will restrict myself in what follows to this simple example: it will be obvious how to extend everything said here to the general case.

If we take a model for a language containing alethic as well as deontic modalities to be a quadruple $\langle W, R_d, R_a, V\rangle$ where W is a non-empty set of possible worlds, R_d the relation of deontic alternativeness, R_a the alethic accessibility relation, and V the assignment function, then we can reformulate – or, rather, specify – definition (2) as follows:

(3) $K = \{O(A), O(B)\}$ is consistent$_3$ iff there is a model $M = \langle W, R_d, R_a, V\rangle$ and a $w \in W$ such that $V(\Diamond(A \wedge B), w) = 1$, i.e., iff there is a model $M = \langle W, R_d, R_a, V\rangle$, and there are $w, w' \in W$ such that wR_aw' and $V(A \wedge B, w') = 1$.

For this simplified case, definition (3), as applied to $K = \{O(A), O(B)\}$, coincides with the following equivalence:

(4) $K = \{O(A), O(B)\}$ is consistent$_4$ iff there is a model $M = \langle W, R_d, R_a, V\rangle$ and a $w' \in W$ such that $V(A, w) = 1$ and $V(B, w) = 1$.

By 'coincidence of (3) and (4)' I mean that it can be shown for every set K that it is consistent$_3$ iff it is consistent$_4$.

My point is that – provided we require of our models that the deontic alternativeness relation R_d is serial – definitions (3) and (4) also coincide with the "normal" definition of consistency of a set of sentences, this "normal" definition, as applied to K, being:

(5) $K = \{O(A), O(B)\}$ is consistent$_5$ iff there is a model $M = \langle W, R_d, R_a, V\rangle$ and a $w \in W$ such that $V(O(A), w) = 1$ and $V(O(B), w) = 1$, i.e., iff there is a model $M = \langle W, R_d, R_a, V\rangle$ and a $w \in W$ such that for every w': if wR_dw' then $V(A, w') = 1$ and $V(B, w') = 1$.

This definition coincides with (4) in the sense that for every set K it can be shown that K is consistent$_4$ iff it is consistent$_5$ because, according to (5), it suffices for the consistency of K to find one model as described, and this can easily be done as soon as we have a model as described in the definiens of (4), and – in virtue of R_d being serial – vice versa.

After having (re-)gained this "normal" concept of consistency for normative languages, we must now explain what we understand by a moral conflict or dilemma. The concept of a moral dilemma as offered by Prof. Barcan Marcus seems to me at least as interesting as her considerations on the concept of consistency. Although she does not give even an informal definition of moral dilemmas as she does with consistency, this definition can be reconstructed from her text easily:

(6) $O(A)$ and $O(B)$ conflict with one another (i.e., they present or involve a moral dilemma) in world $w \in W$, where W is part of a model $M = \langle W, R_d, R_a, V\rangle$, iff $V(O(A), w) = 1$ and $V(O(B), w) = 1$, and $V(\Diamond(A \wedge B), w) = 0$, i.e. iff $V(O(A), w) = 1$, and $V(O(B), w) = 1$, and for every w': if wR_aw' then $V(A \wedge B, w') = 0$.

Now it is easy to prove Prof. Barcan Marcus's claim that the reality of a moral dilemma need not entail an inconsistency: the fact that a moral dilemma is real in a world w, or that two moral principles $O(A)$ and $O(B)$ conflict with one another in a world w, does not imply that there is an inconsistency or that the set $K = \{O(A), O(B)\}$ is inconsistent according to one of the definitions (2), (3), (4), or (5).

According to the standard view, a moral dilemma always implies an inconsistency, but this employs another concept of a moral dilemma. A moral dilemma is usually described as a situation where somebody is under an obligation to do two things as described by sentences A and B whereas it is impossible to do both, i.e., there is no possible world whatsoever where $A \wedge B$ is true. This would give a definition of a moral dilemma like the following one:

(7) $O(A)$ and $O(B)$ conflict with one another (or include a moral dilemma) in world $w \in W$, where W is part of a model $M = \langle W, R_d, R_a, V\rangle$, iff $V(O(A), w) = 1$, $V(O(B), w) = 1$, and for every model $M' = \langle W', R_d', R_a', V'\rangle$ and every $w' \in W'$: $V(A \wedge B, w') = 0$.

A moral conflict or dilemma in the sense of (7) always, of course, entails an inconsistency in the sense described above under (2)–(5).

Whereas according to (7), a moral conflict or dilemma in one world *w* remains a moral conflict or dilemma in every world *w′* with the same moral principles or rules as in *w*, the new view of moral dilemmas as outlined by Prof. Barcan Marcus and defined in (6) is such that we can have a moral dilemma in one world *w* without having it in another world *w′*, although in both, *w* and *w′*, the same moral rules and principles hold.

Appendix 2: Ruth Barcan Marcus, Letter of May 6, 1981, to Edgar Morscher [Commentary on the manuscript of the text reprinted as appendix 1]

Yale University
Department of Philosophy
New Haven, Connecticut 06520

May 6, 1981

Dear Professor Morscher:

I appreciated your letter of April 22 and am pleased that you found my paper interesting.

About my definition of consistency – in proposing it, I was responding to a rather large literature in which it is taken as true or implied that dilemmas signify an inconsistency in a set of moral rules. My definition was originally prompted in response to a symposium paper of John Lemmon's from which I quoted – and who was a logician who did work in modal logic. See also Davidson, Hare and even Ross, where, although unformalized, dilemmas were the impetus for calling rules prima facie (on the supposition of a problem of consistency). Now none of the latter three formally *defines* consistency (except Lemmon), but since the examples they use of dilemmas are precisely the ones I and everyone uses, one supposes that they have in mind a definition of consistency where those dilemmas are a mark of *inconsistency*. One can trace the source to certain principles of deontic logic which are taken to be *valid* and which would generate inconsistency if, in the case of conflicting obligations, there are no principles of priority, qualification clauses and the like. There are a variety of such principles (some explicit in systems of deontic logic, some implicit), which can be seen as the source, singly and in combination, and taken in conjunction with the rest of the semantics (for quantification, the connectives, domain of the variables, etc.).

One such principle is

(A) $O\neg A \rightarrow \neg OA$

Another, taken as an axiom schema for deontic logic as in B. Chellas, *Modal Logic*, Cambridge, 1980, pp. 190–203 (which was published some time after I wrote my paper)

(B) $\neg(OA \wedge O\neg A)$

Another is

(C) $(OA \wedge OB) \rightarrow O(A \wedge B)$

where the latter is taken in conjunction with some version of ought implies can (B or some variant).

I could go on.

Indeed virtually all the usual "axiom" sets for deontic logic would make dilemmas evidence of inconsistency of the deontic systems unless there was patching of rules and priorities.

So, my definition is "new" in one important sense. It does not suppose the validity of standard axioms of deontic logic where formally given (or standard suppositions about such principles where informally given). Such principles *would* render dilemmas evidence of inconsistency.

The definition I propose is as I say in the article a very natural one – and the closest analogue to ordinary definitions of consistency. In the sense of its being a natural and plausible analogue, it is one which is persuasive and should therefore raise questions about principles of deontic logic which have been taken as axiomatic. In your comments you have seen the plausibility of the definition and in that more general sense it is not "new"; that is precisely the point I was making.

I do not believe that I have proposed a new *definition* of moral dilemmas – for if it were a new definition, then my examples of dilemmas would be different from those usually given. But that is not the case. It is rather my focusing on what the source of the dilemma is. The correct analyses ultimately would be a complicated mix of alethic, pure modal, and deontic modalities (and very likely physical modalities as well). There is no one world in which an agent can both return a cache of arms (under a promising rule) and not return it (under a benevolence rule) at the *same time and place in that world*. But there are worlds in which such contingencies would not arise. In any case, we should not suppose $(OA \wedge O\neg A) \rightarrow O(A \wedge \neg A)$ (see above).

In this respect you have seized on an important feature of my paper.

I hope I have answered your questions. With some change in your paper about the sense in which my definition of consistency is "new" (relative to beliefs about valid deontic principles) it would be interesting to publish.

Incidentally, as some evidence of "novelty", the initial response of many quite sophisticated philosophers was that my definition of consistency was "peculiar." They said they would call the rules of my card game "inconsistent." But what underlies the initial response is seeing moral "laws" "rules of games" as analogous to *physical* laws.

With all regards,
Ruth Barcan Marcus
Halleck Professor of Philosophy

References

Barcan Marcus, Ruth (1966): "Iterated Deontic Modalities", in *Mind* 75 (1966), pp. 580–582; reprinted in Barcan Marcus (1993), pp. 40–43.

Barcan Marcus, Ruth (1980): "Moral Dilemmas and Consistency", in: *The Journal of Philosophy* 77 (1980), pp. 121–136 (quoted as 'A' with page number); reprinted in Barcan Marcus (1993), pp. 127–141 (quoted as 'B' with page number) with an introductory note on pp. 125–126.

Barcan Marcus, Ruth (1993): *Modalities. Philosophical Essays*. New York-Oxford: Oxford University Press, 1993.

Barcan Marcus, Ruth (1996): "More about Moral Dilemmas", in Mason (1996), pp. 23–35 (quoted as 'C' with page number).

Belnap, Nuel, and Paul Bartha (1995): "Marcus and the Problem of Nested Deontic Modalities", in Sinnott-Armstrong/Raffman/Asher (1995), pp. 174–197.

Brink, David O. (1996): "Moral Conflict and Its Structure", in Mason (1996), pp. 102–126.

Chellas, Brian F.: *Modal logic. An Introduction*. Cambridge: Cambridge University Press, 1980.

Chisholm, Roderick M. (1963): "Supererogation and Offence: A Conceptual Scheme for Ethics", in *Ratio* 5 (1963), pp. 1–14.

Kanger, Stig (2001): *Collected Papers of Stig Kanger with Essays on his Life and Work*, Vol. I, ed. by Ghita Holmström-Hintikka, Sten Lindström and Rysiek Sliwinski. Dordrecht-Boston-London: Kluwer Academic Publishers, 2001.

Mason, H. E., ed. (1996): *Moral Dilemmas and Moral Theory*. New York-Oxford: Oxford University Press, 1996.

Morscher, Edgar (1982): "Antinomies and Incompatibilities within Normative Languages: Some Semantic Considerations", in Anselmo A. Martino (ed.), *Deontic Logic, Computational Linguistics and Legal Information Systems: Edited Versions of Selected Papers from the International Conference on "Logic, Informatics, Law", Florence, Italy, April 1981*, Vol. II. Am-sterdam-New York-Oxford: North-Holland Publishing Company, 1982, pp. 83–102; see pp.94–97: "Appendix A: Ruth Barcan Marcus on Moral Dilemmas and Consistency".

Morscher, Edgar (2002): "The Definition of Moral Dilemmas: A Logical Confusion and a Clarification", in *Ethical Theory and Moral Practice* 5 (2002), pp. 485–491.

Morscher, Edgar (2012): *Normenlogik. Grundlagen – Systeme – Anwendungen*. Paderborn: mentis Verlag, 2012.

Sinnott-Armstrong, Walter, in collaboration with Diana Raffman and Nicholas Asher, eds. (1995): *Modality, Morality and Belief. Essays in Honor of Ruth Barcan Marcus*. Cambridge-New York-Melbourne: Cambridge University Press, 1995.

Sinnott-Armstrong, Walter (1996): "Moral Dilemmas and Rights", in Mason (1996), pp. 48–65.

Wessels, Ulla (2002): *Die gute Samariterin. Zur Struktur der Supererogation*. Berlin-New York: Walter de Gruyter, 2002.

Michael Frauchiger

Interview with Ruth Barcan Marcus

Frauchiger: Professor Barcan Marcus, from your autobiographical talk during the Symposium, it became apparent that at the beginning of your intellectual journey you found yourself in an area of philosophical and personal tension between W. V. Quine (who tended to set the tone at the time) and C. I. Lewis, whose pioneering work in modal sentential logic was continued by your ground-breaking axiomatization of quantified modal logic. I will turn to the debate about the interpretation of modal logic in a moment, but first I would like to touch on a metaphilosophical aspect of the differences between C. I. Lewis and W. V. Quine concerning Lewis's theory of the imposition of modes of conception and interpretation upon experience—I am thinking of his work *Mind and the World-Order* ...

Ruth Barcan Marcus: Well, the work that influenced me more was *An Analysis of Knowledge and Valuation*, a big book of C. I. Lewis's, and the one that is closest to my own work.

Frauchiger: Oh, I see. Well, actually C. I. Lewis's imposition-view is in some respects similar to Henri Lauener's "open transcendentalism" and other Kantian analytic epistemologies which assume that philosophy is a normative meta-discipline that is *sui generis*, i.e. related to but different from explanatory

---

Interviewer's note: This informal interview with Ruth Barcan Marcus was made on 31 May 2008, the following morning after the 3rd International Lauener Symposium in her honour. We met in the restaurant of her hotel in Bern for an open conversation about her work, the original idea being that the improved transcription of this taped oral interview would be rewritten before publication—but matters should turn out differently. Ruth Barcan Marcus's prolonged, severe illness eventually made it impossible for her to even entertain the idea of addressing the otherwise relevant review of her answers in the interview. I occasionally made mention to her of the planned inclusion of the interview in this book and she never disapproved of it even after it had become clear that she would not be in a position to review it. For Barcan Marcus's frank answers alone, this informal interview is a vivid and valuable document and I am therefore taking the liberty of publishing it in the present volume, in unavoidably condensed and revised form. The revisions I have made in reproducing the oral conversation relate to the linguistic editing of the text where this appeared necessary to ensure readability as well as to the omission of certain passages in the conversation which I deemed repetitive, fragmentary or not directly relevant to the interview's topics.

scientific disciplines. Philosophy understood in such a way, as a methodological meta-discipline, is supposed to provide relative *a priori* foundations—relative, that is, to current practical situations, interests, values, conventions. Epistemology *qua* methodology, then, provides *post festum* foundations for other scientific disciplines, reconstructing their objectives and systematizing their conceptual and empirical means, thereby endeavouring to justify their epistemic adequacy and reliability.

On the other hand, in contrast to such Kantian approaches, W. V. Quine—and Ernest Nagel, for that matter—reject any requirement to secure independently some warrant for the methods of empirical science and insist that useful philosophy is to be seen as a very abstract descriptive discipline that is seamless with empirical science itself and is thus not to be seen as a methodological meta-discipline. – Now in your 1993 book *Modalities*, you write that, at Yale, the profound philosopher (as you call him) Ernst Cassirer succeeded to direct your philosophical predilections into some Kantian channels. Do you think of philosophy as a separate, methodological discipline or rather as a collection of very abstract inquiries that are seamless with empirical research itself?

Ruth Barcan Marcus: Well, seamless is ambiguous. Kant begins, the way I read him, with the question: what makes science possible? He assumes that there is such a thing as science—that there are true scientific claims. And how can you explain that on purely empirical grounds? I mean, Hume has challenged that and we have to answer to Hume. He asked the same question in ethics: what makes morality possible? And the answers go along similar lines. The only, major difference, though, is that there is this understanding of Kant that we are not free to choose any ground whatever. There are the "Dinge an sich", and I always viewed them as limiting not our capacity but our choices when we are doing science. We can't decide that a body has a maximum velocity of 186.000 miles a second or whatever. That's not up to us, but we have structured it. It's the world out there which we can't describe directly, which keeps us honest, so to speak, and makes it possible for us to do empirical science. And that's where I was influenced by Kant, by these questions of what makes science possible, what makes ethics possible. He answers them in similar ways but also in different ways, for example he doesn't answer the question in the same way with respect to religion; that's a wholly open question to which there is no analogous answer like what makes religion possible. – I also was influenced by Cassirer because of the emphasis he places on the sciences as did Kant. And in that respect, there is a similarity between the positivists (although Cassirer would have denied it) and views like Cassirer's because central is an analysis of science. That's the naturalism. And I don't think of Kant as juxtaposed against naturalism.

Frauchiger: But would you say that philosophy can be considered as a kind of a methodological meta-discipline? Or would you rather say it's part and parcel of empirical science, just a very abstract section of it?

Ruth Barcan Marcus: Well, a lot of it, I think, is a meta-discipline. I mean the very question of here we have science, now we ask a meta-question: what makes it possible? That's a methodological question. That's not a question that can be answered by the physical sciences directly.

Frauchiger: I'd now like to proceed to another main difference between Lewis and Quine at the time, which concerns the foundation and interpretation of modal logic.

Ruth Barcan Marcus: Yes. Quine attributed certain characteristics to Lewis, which were not justified. Although sometimes—if you read Lewis in a very sketchy way, which apparently Quine did—it looks like he's just looking for some language which will capture deducibility, necessary consequence, and that's why he introduces the strict implication relationship and introduces it in a way which muddles up use and mention. But as you see in the essay that I wrote on Quine's changes of mind over time[1], C. I. Lewis used the metaphoric possible worlds to give some meaning to the modal operators. He wasn't just interested in getting some symbol to capture deducibility. There are several quotations—I used one in which Lewis says: "You know, in this world I have so many coins in my pocket, there might be a world in which I had fewer coins in my pocket." You think it was Kripke talking. (smiling)

---

1 ***Explanatory note:*** Barcan Marcus is referring to her essay "A Backward Look at Quine's Animadversions on Modalities" (in Ruth Barcan Marcus, *Modalities: Philosophical Essays*, New York, Oxford: Oxford University Press, 1993, 215–232).

Frauchiger: That's interesting. When did C. I. Lewis formulate this rough anticipation of the idea of possible-world semantics—long before Carnap started to write about state-descriptions?

Ruth Barcan Marcus: Well, he didn't work out a semantics but apparently it was implicit because if he was challenged about the modalities he said: "Well, you know, we can talk about possible worlds in which things that don't happen in this one might have happened in the other one." But it's not a full-fledged semantics—as a matter of fact, full-fledged semantics wasn't around when Lewis started out.

Frauchiger: Incidentally, during the discussion at the symposium, you mentioned that Lewis and Quine had both philosophical and personal differences ...

Ruth Barcan Marcus: Well, I'm saying something tangential: there's a very well known literary critic at Yale named Harold Bloom. And he wrote a book called—or he has a thesis about—"The Anxiety of Influence", which is about the tendency of younger people to revolt against the established figures in their field. And so that may not be just Lewis and Quine but a tendency of younger scholars to try to establish their place in the hierarchy, and that may require rejecting the established figures. But it was more than that—I mean, I'm only speculating about the details, but what was more obvious was that Quine had travelled all over Europe and talked to all the positivists. And Lewis was not a positivist and he objected to much that was then said to be a characteristic of positivism. And so, I think, Quine thought that Lewis was hampering intellectual progress—partly it was that he dominated the department and that Quine really thought that his logic was not of any interest, that it was just a big mistake from beginning to end ...

Frauchiger: That is, Quine thought modal logic to be based on a use-mention confusion, as regards the (alleged) analysis of logical consequence (and other semantic notions) at the object level in terms of strict implication ...

Ruth Barcan Marcus: Right. But you'll notice in my paper on Quine[2] that Carnap often uses 'implication' in the way that Quine disapproved of. ... You see, it's the difference between an object-language symbol and the relation of consequence, but what I point out in that paper is that it was not written in stone how to use the word 'implication' and whether it should always be used as distinguished from consequences. That seems to me a superficial concern ... and if you read my paper you will see that there was very good precedent. – However—this went beyond Quine—one of the things I showed is that in Lewis's favorite system, S3, you couldn't prove a classical deduction theorem. And so it seems peculiar to say that you have captured this relationship when you can't even prove the deduction theorem. (But actually in S4, you can prove a version of the deduction theorem which I think is more intuitive ... in S4 you can prove what is the best version of the deduction theorem.) – So the notion that the motivation in Lewis's ... that the systems that he developed are all questionable, full of confusion, is not very well supported.

Frauchiger: So Quine started his attack on the foundations of modal logic by objecting to Lewis's systems. And then, when you published your first two papers on quantified modal logic ...

Ruth Barcan Marcus: That got him started again—you always almost had the feeling he was waiting to pounce. As soon as it came out, he wrote that paper. He didn't wait a minute. Same with the identity paper[3]: as soon as it came out, he wrote a bad review. That's why I said it was like the boy trying to stop the leaks of the dike. As soon as he thought he had disposed of modal logic, along comes somebody who does quantified modal logic—even more preposterous!

2 See footnote 1.

3 ***Explanatory note:*** Barcan Marcus is referring to her paper (signed with her birth name 'Ruth C. Barcan') "The Identity of Individuals in a Strict Functional Calculus of Second Order" (in *The Journal of Symbolic* Logic 12/1, 1947, 12–15). – Not long beforehand, she had published two other papers on quantified modal logic, viz. Ruth C. Barcan, "A Functional Calculus of First Order Based on Strict Implication" (in *The Journal of Symbolic* Logic 11/1, 1946, 1–16) and Ruth C. Barcan, "The Deduction Theorem in a Functional Calculus of First Order Based on Strict Implication" (in *The Journal of Symbolic* Logic 11/4, 1946, 115–118).

Frauchiger: So Quine intensified his attack by claiming that your systems involving quantification into modal contexts inevitably lead to deep problems of interpretation.

Ruth Barcan Marcus: Inescapably, to intractable problems ... (smiling)

Frauchiger: But the interesting thing is that this was actually the beginning of decades of debate on modal semantics and ontology.

Ruth Barcan Marcus: Well, that's because everybody went along with him, except for Lewis up to that time. They just figured he knew what he was talking about.

Frauchiger: Still, finally even some of his own students, particularly Dagfinn Føllesdal in his 1961 dissertation, were beginning to move away from Quine's notorious doubts about modalities.

Ruth Barcan Marcus: Well, it depends. Dagfinn wrote a dissertation, and then he sent around a copy of his dissertation which was changed from the original dissertation. And in the original dissertation he was much more critical. But then, when he rewrote his dissertation, he became more tolerant.

Frauchiger: Even if so, the rewritten version of his dissertation was published in 1966, which is not very much later.

Ruth Barcan Marcus: Not very much ...

Frauchiger: Yet Quine kept on insisting on his various doubts about modalities and modal logic.

Ruth Barcan Marcus: He never gave up on them. Let's put it this way: he had a lot of reservation and little by little it turned out that he was wrong or mislead or made mistakes. So in the end he keeps talking about "you're committed to essentialism." And not only that: "You're committed to Aristotelian essentialism." Nothing was more remote from Aristotelian essentialism than what he was talking about. But after I gave this analysis of what essentialism might be like, mine was Aristotelian essentialism: Aristotle has no essential entities. A lot of people are talking about essential things, like your essence and my essence. That's very Carnapian and very metaphysical ... and very, in a way, Cartesian: like your essence is your soul or whatever; and your essence doesn't get shifted from world to world, your essence is rigid. And I think even somebody like Kit Fine tried to give an account of individual essence. That is not what Aristotle was focusing on or what I was—I was talking about essential properties. Aristotle said essentialism was about essential properties. And he kept saying that modal logic is committed to essential properties. Well, in the end, what modal logic is committed to is talking about essential properties; even if you want to say there are none. The way you would deny essentialism is to proof that given that this is the structure of an essential sentence, there are modal systems consistent with every one of them being false. And that was finally proved. I just speculated—I said, I can't think of any of these claims that any modal system that I'm familiar with would hold true, particularly people's favorite modal systems like S5. And then, my then colleague Terry Parsons came along and did a very careful proof to show that there are systems of modal logic in which every essentialist modal claim is false.

Frauchiger: In retrospect, what do you think was the outcome of this long-winded debate?

Ruth Barcan Marcus: Well, what was the outcome? People began to do modal logic. There is now a society of modal logic. There are now publications called: "Recent Advances in Modal Logic". I mean, it created a cottage industry, and among people who are quite respectable, many of them have positions. So the idea that it was nonsensical or committed you to nonsense—I mean, that's how he starts out that original paper: he's going to show how this introduction of quantifiers into modal logic is a terrible muddle. That's all gone. I think a lot of the stuff that's being done in modal logic is probably not the best logic in the world or not the most interesting, but there it is. And some of it is interesting. A lot of it is interesting.

Frauchiger: It is also interesting to note that your work has been a crucial factor in the widespread revival of intensional formal theories ever since the 1960s. And what I find particularly remarkable is that you have decisively further clarified the concept of extensionality. In this way, Quine's too austere concept of extensionality, which originates from his predilection for systems in the form of his canonical notation, is overcome by your distinction between different grades of extensionalizability and intensionality.

Ruth Barcan Marcus: Yes, what you notice in what I propose is that the idea of extensionality is tied to the language that you are using. For example in my systems of modal logic, if you go to second order, it has at least three interesting equivalence relations: the triple bar, the quadruple bar and then identity at the lowest level. I don't introduce, although I noticed that Tim [Williamson] was talking about, identity of properties. I allowed quantification over them, that's how I defined identity for individuals, but he's going to extend identity to properties (properties are identical if they have all their properties in common). So a language is extensional, I proposed, to the extent that it reduces a stronger to a weaker equivalence. For example in *Principia Mathematica*, Russell says that we can use two equivalent ... —for him descriptions are like properties, they are equivalent—as if they were identities, if we stick to mathematics; we don't need this distinction if our language is only going to include mathematical discussions. So he says that in *Principia Mathematica*—whereas in "On Denoting" what he points out is that if you're going to extend your language beyond that, then you have to have a language in which triple bar equivalences are not called identities because of the substitution problem in belief contexts and so on.

Frauchiger: I think yours was a very novel clarification of the concept of extensionality ...

Ruth Barcan Marcus: Yes, and it hasn't been ... somebody named Anderson has discussed it ... but even though it came out in this collection of Linsky's, which is very widely read, it wasn't seized upon, somehow.[4] Have you noticed that it has been taken up?

4 ***Explanatory note:*** Barcan Marcus is referring to her paper "Extensionality" (in *Mind*, n.s., LXIX, 1960, 55–62; also reprinted in Leonard Linsky, ed., *Reference and Modality*, Oxford: Oxford University Press, 1971, 44–51). – Barcan Marcus's paper "Extensionality" overlaps to some extent her path-breaking 1961 paper "Modalities and Intensional Languages" (in Ruth Barcan Marcus, *Modalities: Philosophical Essays*, New York, Oxford: Oxford University Press, 1993, 3–23).

Frauchiger: No, surprisingly enough, not, even though there are still many people who are interested in extensionalism of some kind but who do not want to hold on to Quine's austere concept of extensionality. And I think your distinction between different grades of intensionality and extensionalizability, invoking as criteria for the distinction of grades the diverse types of characterizing identity, should certainly be very appealing for many of them.

Ruth Barcan Marcus: I'm waiting for that to be more widely influential. I keep expecting it will be. ... I use it in a way, but e.g. in my paper on attributes and classes[5], I couldn't because there is no good vocabulary for it—so I didn't know whether to use 'collections' or 'sets' and 'attributes'—and because the discussion of extensionality is misleading: what does having the same members come to? An evening star and a morning star have the same members but the attributes aren't identical, it's just that the members are the one member. And the interesting thing is: if you are confining yourself to collections, itemizations of individual things named by proper names, if they are the same collection, then they are necessarily the same collection. That follows directly from the necessity of identity because each of those items has an identical reflection in the rearranged list of names. And so the necessity applies to each of them and so it applies straightforwardly to the whole. So you don't need an axiom of extensionality. The only time you need an axiom of extensionality is when you want to reduce some other kind of equivalence to some lesser kind of equivalence, if you want to reduce modal equivalence to just material equivalence. Now, one of the things that didn't get noticed, and still isn't explicitly noticed, is that all the while, everybody kept saying you have to have these restrictions on scope—which you do—but they made it sound as if it were arbitrary: you're not supposed to have this kind of expressions in the scope of a modal operator. The fact is that it was provable in my quantified modal logic: it was an interesting, complicated meta-theorem about substitution that you cannot substitute within a modal context lesser forms of equivalence like triple-bar equivalence. You can always substitute with respect to lowest level objects' identity. Which raises the question which is part of the literature about belief: it looks as if in belief contexts, you can't even do that. But that's another story. So more strongly intensional contexts than modal contexts may still present a problem.

5 ***Explanatory note:*** Barcan Marcus is referring to her paper "Classes, Collections, Assortments, and Individuals" (in Ruth Barcan Marcus, *Modalities: Philosophical Essays*, New York, Oxford: Oxford University Press, 1993, 89–100).

Frauchiger: Thus, from the novel point of view that you introduced, not only languages based on classical logic but also many modal systems must be regarded as extensionalized languages though they are of course more intensional and less strongly extensionalizable than classical languages.

Ruth Barcan Marcus: Yes, because of what we just said before: there is some reduction. But belief contexts are omitted.

Frauchiger: They are explicitly intensional.

Ruth Barcan Marcus: Yes. Well, so most people would regard modal contexts as intensional. But they're very different because the intensional contexts like belief involve knowledge attitudes, which modal logic doesn't. So sometimes people decide that 'intensional' means you can't substitute these weaker forms of equivalence. But there are these degrees of intensionality. And in a language—if it's possible, I don't know if it's possible—which formalizes belief contexts, to be able to substitute modal equivalences in belief contexts would be an extensionalizing principle.

Frauchiger: Indeed. However, some authors working in modal metaphysics show no concern at all about extensionality—I'm thinking of David Lewis and adherents of modal realism.

Ruth Barcan Marcus: For David Lewis, the whole idea of the necessity of identity doesn't make very much sense. If David Lewis got to talk about possible worlds, there is no necessity of identity. The name of an object doesn't migrate unequivocally to another world. There might be something in that world that's very much like this one, but it's a very different structure. But I don't know—do you regard David Lewis as a modal metaphysician? There are modal metaphysicians, but I never classed him with the modal metaphysicians. But maybe I should, I don't know.

Frauchiger: Returning to your own approach to modal semantics: you have suggested a substitutional semantics for certain special kinds of discourse. On the other hand, as early as 1961, at the end of your paper "Modalities and Intensional Languages", you presented a semantics with objectual quantification for intensional languages ...

Ruth Barcan Marcus: Where? This semantics at the end of the appendix?

Frauchiger: That's what I mean, at the end of the paper ...

Ruth Barcan Marcus: It's not for all intensional languages. What I did was present a model-theoretic semantics which satisfies the Barcan formula. It fits S5—S5 is not my favorite modal system—but all I wanted to do there was to show that there are plausible semantics for modal logic: here is an example.

Frauchiger: Yes, and since this model-theoretic semantics validates the Barcan formula, it is not committed to possibilia correlated with intensional objects. So Quine's "flight from intensions" is unnecessary regarding your semantic approach.

Ruth Barcan Marcus: Of course, that was his criticism of Carnap: that you needed this duality intension and extension and in all modal contexts it was an intensional object and in all non-modal contexts it was an extensional object. He just attributed that to me because we were both doing modal logic, without examination. And, even though I presented this model-theoretic account which showed that you could have a reasonable semantics for modal systems, he didn't pay any attention, neither did anybody at that time.

Frauchiger: Not even Kripke?

Ruth Barcan Marcus: No ... he never acknowledged it. I mean ... for example, when I talk incidentally about proper names, I say in that paper that I didn't get this idea just because I needed proper names—at the time, this is pre-Chomsky in linguistics, they said: proper names have no meaning; they're not lexical items; they did not put them ordinarily in dictionaries, they sometimes put them in biographical dictionaries where the dictionary functions as an encyclopedia instead of as a lexicon, although later they would pick out some descriptions and include them in the body of the text; but I remember being told by a linguist that proper names are not lexical items. And incidentally, when I said they're like tags, they only have reference, they don't have lexical meaning, everyone says: "Well, that's like Mill." Actually it's quite different from Mill, but nobody bothers to go into Mill and see exactly what Mill said (smiling). They just keep repeating these mistakes. Anyway, when Quine keeps talking in the "Discussion"[6] about essentialism and says that proper names are a red herring, Kripke rises to the occasion and says that maybe the idea of proper names is what essentialism is about—that a thing has its name necessarily. That's not what I said although it's an interesting idea and since Kripke is so religious, it's like Adam who knew the names of things (I think it was Adam, in the old testament). As if the names of things were essential, God-given. And so I remember when I later on talked to Dagfinn [Føllesdal] what a misunderstanding that was, Dagfinn said: "Oh, maybe he was just trying to explain what Quine said." But that's not the way it came out—he thought he was adding to an understanding of what was going on by saying that essentialism can be understood as things have their names necessarily.

Frauchiger: Yes, and your exclamation, "Oh, no." towards the end of the "Discussion", occurs after the first paragraph of Quine's response to Kripke, but it actually applies much more to Kripke's statement.

Ruth Barcan Marcus: Sure, well it applies to Kripke because he thinks he's explaining what is meant by essentialism. – It's really a pity that Russell gave up so easily because when you read early Russell, like "Scott is the author of Waverly", he says: "Oh, well, George IV isn't interested in knowing whether Scott was Scott." 'Scott' just refers to Scott, it's just whether this descriptive account is true. But then he gave all that up because he didn't give an account (even though the linguists didn't worry then very much) of how it was possible for someone who has a confrontation with an object like Bismarck—when he uses 'Bismarck', he's referring to the object—but if he doesn't encounter Bismarck, many eons later, how can he be talking about Bismarck (because he

6 ***Explanatory note:*** Barcan Marcus is referring to the transcribed and edited "Discussion" which followed her presentation of "Modalities and Intensional Languages" in February 1962 at the Boston Colloquium for the Philosophy of Science. That "Discussion" is included in her 1993 collection of essays as an Appendix to her paper "Modalities and Intensional Languages" (see Ruth Barcan Marcus, *Modalities: Philosophical Essays*, New York, Oxford: Oxford University Press, 1993, 24–35).

hasn't encountered him)? That was a big problem for Russell—and for me, but I was not hasty about giving up this notion of direct reference because I couldn't explain this phenomenon, which was then explained, not by Kripke although he takes credit, but by Geach. The way that happens is: somebody saw Bismarck. What Geach is saying: at the outset, there has to be the object there to be named; and then this name is carried along historically in our language and when we use it, even though we may not know this path, we are referring to that same thing that was referred to all along this historical chain. So: I dug my heels in and said, "There's got to be an answer to this question." Russell was too impatient, although he was always a little reluctant to give up the special role of proper names. If you read Russell, 'Scott' is never reduced to a description in the way that 'the author of Waverly' is a description, but he in the end decided that maybe proper names are things like 'this' or 'that'. But then they of course won't carry into the language—they are like a private language because they disappear. So I just hoped somebody like Geach would come along, and in Finland [at the Colloquium on Modal and Many-Valued Logics in August 1962 in Helsinki] Geach was there and he already expressed this view, although it wasn't published until 1969—i.e. before Kripke—in "The Perils of Pauline", which then got incorporated in 1972 into a book [Peter Geach, *Logic Matters*]; it was published and in that eloquent way, you know: "... in apostolic succession down to our own times". ... And people like Soames—I mean, Soames is smart, but historically he has done the history of the subject a disservice—he says: "Kripke's whole idea is very simple: you dub a thing, you call it by a name and then it gets carried along ..."—that was almost exactly what Geach says. Similarly the necessity of identity: he was inclined to think it was Kripke's original idea.

Frauchiger: Since the end of the 1940s, you have argued that Quine's notorious examples of substitution failures in modal contexts can be successfully dispelled within interpreted quantified modal logic.

Ruth Barcan Marcus: Oh, surely. Gupta repeats all of that later on, about how I dispelled all of those.

Frauchiger: And you have insisted that genuine proper names are not what was later called "rigid designators" by Kripke.

Ruth Barcan Marcus: No, although they have features in common. 'The successor of 2', for example, is a rigid description: if something is the successor of 2, it's the successor of 2 in every alternative world. So it shares that property—but it has structure—and the important thing about proper names is, they have no structure, they're just tags.

Frauchiger: In contrast to rigid descriptions genuine proper names are tags, which have no meaning, no semantic content ...

Ruth Barcan Marcus: They just refer, that's right.

Frauchiger: And later on, you augmented this direct reference view of proper names by adopting Donnellan's causal or historical-chain account of naming ...

Ruth Barcan Marcus: Well, by that time I believed in the historical-chain account. And I liked Donnellan's paper because if you use the historical chain and it goes back to nothing, then there is no Jacob Horn, this supposed American who was born in Washington, Pennsylvania, about whom this hoax [the *Horn Papers*] was written. And this isn't speaking about nothing. You trace it. And I said: suppose someone believes it all and wants to do research, find out more about this person, and then he isn't there, there is no such person. – But that was after the historical chain was in place.

Frauchiger: In your 1986 paper "Possibilia and Possible Worlds", you write that on a historical-chain account of direct reference, genuine proper names have an ostensive origin, are traceable back historically to the publicly given and named object.

Ruth Barcan Marcus: Yes, in 1976 I gave a Presidential Address about possibilia, and there I coined the phrase—which other people have used—that proper names are like a long finger of ostension. I like that.

Frauchiger: But this thesis of ostensive givenness appears to imply that a group of people can directly and absolutely—perhaps even pre-linguistically—be linked with one and the same object as a result of somebody merely pointing to it in front of them ...

Ruth Barcan Marcus: Well, yes. Davidson raises these questions. I don't know if that is what you are referring to.

Frauchiger: Well, I think of the Quinean objection that the practice of ostension does not on any account guarantee uniform reference.

Ruth Barcan Marcus: It doesn't *guarantee* it because at a christening, something might go wrong. But if everything is carried out according to the allowed practices, then it's very likely that nothing is going to go wrong. But in the Davidson case, you don't point to somebody—because let's say you're polite, you're not allowed to point—so you say: "The man over there drinking a Martini ... " and so you have directed the attention of somebody instead of pointing to somebody, but it's not crucial that he is drinking a Martini. Somewhere I said that's very much like a throwaway proper name. Let's call him "the man drinking the Martini", even though he isn't drinking a Martini. It's in lieu of ostension. So that is a partial explanation of how this can happen, that is, how you can locate somebody in your empirical field, using a description which is a misdescription. But you didn't intend it, that wasn't your purpose, you just wanted to direct attention to that thing the way you do with a name.

Frauchiger: Still, returning to the basic act of ostension, I wonder if ostension can reliably ensure that a group of people is given, or faced with, one and the same definite object. For if I point to over there, I might be pointing to a concrete object or to a colour. You have no way of knowing what precisely I am referring to ...

Ruth Barcan Marcus: The usefulness of names is that you have an object you think of as an object ...

Frauchiger: As a concrete object?

Ruth Barcan Marcus: As a complete object ... never mind that you're focusing on a color, somebody else is focusing on a shape, but you're referring to the whole object. There could be, I suppose, confusion, when I'm just pointing at the color. But in late Russell, when he says, if you want to think of an individual as a concatenation of properties, you can't point to each of the properties ... nobody is going to ... you're going to hook on to it, like with a tag. If I tag something and you tag something, you are maybe focusing on different aspects of that thing than I am, but we know what it is to tag the whole thing.

Frauchiger: But if we take Quine's "Gavagai" examples ...

Ruth Barcan Marcus: Oh God, yes.

Frauchiger: The native might be referring to the present rabbit or generally to rabbits or to rabbit parts or to the always accompanying rabbit-flies ...

Ruth Barcan Marcus: I think ... that's never been an example for me. If you and I name something—I suppose it's possible that we're naming a part or a feature—but it's usually in the context: if we are talking about a physical object, it has structure, location, it has parameters that contain it and in fact, we can make discoveries about what we have named. Oh—let's call the name 'Tom'—Tom is green, and somebody else says Tom is tall, but the fact that we've tagged this object is enough for us to use that name directly.

Frauchiger: So for ostension to facilitate naming we need to specify a context, a practical situation.

Ruth Barcan Marcus: Well, we usually have to specify a way we arrive at names. Some kind of procedure for naming like a christening or a dubbing or a "let's call that thing so and so".

Frauchiger: So once we have deliberated about what our focuses and aims are in the naming to be done, then we can go on and use ostension?

Ruth Barcan Marcus: Yes, we can. I was never persuaded by the "Gavagai" examples. I don't think the likelihood of that kind of confusion is very strong. We recognize a rabbit as a rabbit. Why would we be referring to the rabbit parts unless there was some reason for that in a certain context, biological, historical—but we have a sense of what the whole thing is. Were you persuaded by those examples? They are psychologically unpersuasive, let's put it that way.

Frauchiger: I was persuaded by those examples only in as far as they show that we can't simply point to a clear-cut object in "the" external world—I doubt that it is possible to comprehend such a mere pointing. But if we manage to intersubjectively describe the relevant context and then go on and use ostension, then I think you're right, ostension works.

Ruth Barcan Marcus: Well, sometimes we don't have to describe the context. We know the context, I mean, we're in it together. When you and I go to a christening, we know what the context is and so we know that we are giving a name to that baby—we're not naming its knees or its elbows.

Frauchiger: Yes, for sure. (smiling) – One final question: more recently, you have rejected language-centered theories of belief ...

Ruth Barcan Marcus: I've always rejected them. I just didn't write about it. (smiling)

Frauchiger: Yes—inclusively you reject quasi-linguistic accounts according to which beliefs are e.g. relations to inner sentences in Fodor's language of thought.

Ruth Barcan Marcus: Yes. Well, if you really believe that you need language to have beliefs—that's what Davidson argues—then you have to say non-language-users don't have beliefs. As I said to Fodor: "Well, what about the very good evidence that higher animals have beliefs—you have all these examples of Dennett's and psychologists'?" He said: "Well then they must have a language of thought." And it occurred to me that when I was in grade school, I think, there was this theory that chimpanzees could talk, they just didn't have the right vocal chords. If they had the right vocal chords, they'd be able—well, that's essentially something like what Fodor was saying. I said: "Well, that takes me back to the theory that if chimps had the right vocal chords they could transform the language of thought into a familiar language." It would seem to be that way. So there's this feeling that language was so crucial to an understanding of belief or knowledge—any of these epistemological claims that it had to be there, somewhere, somehow. And then you can't explain a lot of behavior. Of course animals won't have beliefs about language if they don't have language. And they might not be able to have these higher order beliefs about deception—and Davidson says if you don't have the concept of a mistake you can't claim to know anything or to believe anything—which is ridiculous: animals and human beings know when they're mistaken, but very few of them have the conscious concept of a mistake. You ask anybody in the street what a mistake is and they'll have trouble telling you, but they can recognize it. When I said they probably don't have beliefs about deception, all this is purely speculative, Dennett produced all these examples of dogs that deceive, how they successfully deceive. But there are some people who are so unwilling to accept that possibility that—however persuasive an example you produce—they will say: "But then they're just like a rocket or a thermostat." So, if you don't have a language, you're a rocket or a thermostat.

Frauchiger: As an alternative to such language-centered accounts of belief, you propose a conception of beliefs as attitudes to states of affairs ...

Ruth Barcan Marcus: Right, because I want to get the word truth out of it. I'm very Tarskian. Truth is—and is generally used as—a feature of a linguistic claim. The disquotation principle is all about assenting to sentences.

Frauchiger: But if you construe an agent's beliefs as relations to states of affairs, how do you conceive of such states of affairs?

Ruth Barcan Marcus: I don't—and that's a problem ...

Frauchiger: Are they akin to Russellian singular propositions?

Ruth Barcan Marcus: Yes. When Russell tells Frege that the mountains and valleys, etc., are in the proposition—that's what I'm talking about (whereas most people use the word 'proposition' talking about a sentence, a linguistic entity). That's the beginning of an understanding of what a state of affairs is. The people who have this linguistic view think that you always have to have the sentence before you before you acknowledge the object. This example of the dog and the master in the desert seeing a mirage—what is supposed to be happening? They are saying: "Oh, there is water, there is water, there is ... "—except that the human being could say that, but the dog can't. But they're behaving exactly the same way: they're both thirsty; they're panting; they see the mirage; they rush towards it; there is no water there; they behave in ways that show that they were mistaken. One of them doesn't have language—so he's an automaton and the human being is not? Besides which, it is very likely that even the human being is not always having this sentence before him: "Oh, there is water, there is water, there is ... " (smiling)

Frauchiger: That's well, but the problem about such Russellian structures or singular propositions is that they include at the same time concrete, material objects and abstract, platonic properties and relations ...

Ruth Barcan Marcus: Yes, they have these connections ...

Frauchiger: ... and thus appear to be strangely heterogeneous entities.

Ruth Barcan Marcus: I'm not on top of that. Of course there are all these people who write about it, like Armstrong. I don't know if he has succeeded in giving a good account of states of affairs. But on the whole it's much more promising and persuasive than thinking that sentences always have to be present before you claim to know or believe, or act as if you know or believe.

Frauchiger: I guess you would agree, though, that such states of affairs exist in specific domains and are not simply, i.e. without involving belief, apprehended in "the" independent external world. We do not directly have access or refer to states of affairs: they are out there and have nothing linguistic about them, but their existence needs to be relativized to specific practical contexts of inquiry.

Ruth Barcan Marcus: Well, [there's] the fact that I recognize states of affairs—I'm presumably able to see something of their structure, even the non-language-user does—but I think there's a lot we have to know about human psychology before I can be very sure, or you can be very sure exactly what's going on there.

And I think that's now widely accepted, that not all epistemological circumstances are tied to language.

Frauchiger: But still, I guess that although such states of affairs are not themselves linguistic or quasi-linguistic entities, they are (conceptually) described states of affairs.

Ruth Barcan Marcus: Well, we don't necessarily describe them. And sometimes we're not very good at describing them.

Frauchiger: But they are not simply, directly given to us (I think of C. I. Lewis's imposition-view).

Ruth Barcan Marcus: No, we structure our experience.

Frauchiger: But doesn't that mean that we structure these states of affairs ultimately by means of language?

Ruth Barcan Marcus: Well, is language the only way of structuring them? I'm not sure. I mean non-language-users must be structuring—it's not just a big blur as far as they're concerned. I mean, something's going on which is like: "Oh, there's water over there." But it isn't linguistic: it's recognition of some sort—like I point out in the case of a non-language-using infant: it hears familiar footsteps; it will behave in a way it behaves if it feels his mother is coming, but there's no language involved. There's structuring something: they're separating the sound of the steps. Maybe it's pictorial, I don't know. I mean, it's the business of psychologists who actually are working on these things now.

Frauchiger: Professor Barcan Marcus, thank you very much indeed for this interview.

Contributors

Ruth Barcan Marcus (1921–2012) was Reuben Post Halleck Professor Emerita of Philosophy at Yale University. Her pioneering and influential publications include *Modalities: Philosophical Essays* (1993) as well as numerous academic articles, mainly on the axiomatization and interpretation of quantified modal logic (inter alia on the Barcan formula, the necessity of identity, and second-order modal logic), on Philosophy of Language (notably on proper names and the theory of direct reference, on the interpretation of quantification, and on extensionality and intensional languages), Metaphysics (especially on identity and things, possibilia, classes and attributes, and essentialism), on Epistemology (particularly on epistemological attitudes, belief, and rationality), Moral Philosophy (principally on moral conflict) as well as on some historical figures in philosophy (such as Russell). Barcan Marcus served as President of the International Institute of Philosophy, as President of the Association for Symbolic Logic, and as Chair of the Board of Officers of the American Philosophical Association. She was a Fellow of the American Academy of Arts and Sciences. Barcan Marcus was a recipient of various prizes and awards, including the Medal of the Collège de France, the Quinn Prize (for service to philosophy), and the Lauener Prize for an Outstanding Oeuvre.

Pascal Engel has taught in universities in France (Grenoble, Caen, Sorbonne) and in Switzerland (Geneva) and has held a number of visiting posts. He is presently Directeur d'études at the École des hautes études en sciences sociales in Paris. He has written on the philosophy of logic, philosophy of language and philosophy of mind. His main current work is on issues in epistemology, mostly truth, belief and norms. He is the author, among other works, of *The Norm of Truth* (1991), *Davidson et la philosophie du langage* (1994), *Philosophie et psychologie* (1996), *La dispute* (1997), *Truth* (2002), *Ramsey, Truth and Success* (with Jérôme Dokic, 2002), *What's the Use of Truth* (with R. Rorty, 2007), *Va savoir* (2007), *Les lois de l'esprit* (2012). He has been editor of Dialectica and is a member of the International Institute of Philosophy and of Academia Europaea.

Dagfinn Føllesdal (b. 1932), Clarence Irving Lewis Professor of Philosophy at Stanford University and Professor emeritus at the University of Oslo. After studying mathematics and science in Oslo and in Göttingen he earned his Ph.D. at Harvard in 1961. He taught there until he returned to Norway to teach there and at Stanford University from 1966. He has written or edited around 25 books and 200 articles, mainly on Philosophy of Language, Philosophy of Science and the Humanities, and 20th Century Philosophy. He is a Member of the American Academy of Arts and Sciences and various other academies and has served as President of the Norwegian Academy of Science and Letters. He has received various prizes and awards, including the Lauener Prize 2006.

Michael Frauchiger, Dr. phil., has been lecturing in the fields of Philosophy and Methodology of Science and the Humanities with the University of Applied Sciences of Zurich, the Open University and the University of Bern, and additionally has been an Advanced Research Fellow of the Swiss National Science Foundation. Since its establishment in 2003, he has been the Managing member of the board of trustees of the Lauener Foundation for Analytical Philosophy. His areas of research are Epistemology, Philosophy of Language and Logics, Ontology as well as some interrelated topics in Ethics and Philosophy of Psychology and Action.

Edgar Morscher, born 1941 in Bludenz, Austria, studied at the University of Innsbruck (Ph.D. 1969), *Habilitation* in Philosophy 1974. Full Professor of Philosophy at the University of Salzburg 1979–2009, Chairman of the Philosophy Department 1981–84 and again 2001–09, Director of the Research Institute for Applied Ethics 1984–2001 and 2009–12, Visiting Professor (i.a.) at the University of California, Irvine (1975–76), and at Stanford University (1988–89). Dean of Humanities (1990–92), Rector (1993–95) and Vice Rector (1995–99) of the University of Salzburg. President of the Austrian Philosophical Association (1992–94). Research and teaching interests: ethics, ontology, philosophical logic and semantics, philosophy of the 19th and 20th Centuries, contemporary philosophy. Approximately 400 publications.

Erik J. Olsson is Professor and Chair in Theoretical Philosophy at Lund University, Sweden. His areas of research include Epistemology, Philosophical Logic, Pragmatism, and, more recently, Philosophy of the Internet. Olsson has contributed numerous books and articles to the subjects of epistemic coherence, reliabilism, social epistemology, logic of belief revision and American pragmatism. Recent books include *Against Coherence: Truth, Probability, and Justification* (Oxford University Press, 2005), *Knowledge and Inquiry: Essays on the Pragmatism of Isaac Levi* (Cambridge University Press, 2006) and *Belief Revision Meets Philosophy of Science* (with Sebastian Enqvist, Springer, 2011).

Joëlle Proust is a French philosopher working at École Normale Supérieure, Paris. A researcher at CNRS since 1976, she initially studied the history and the philosophy of logic. (*Questions of form*, 1989, was awarded the bronze medal of CNRS). Since then, her main interest has been in the philosophy of mind: intentionality and animal cognition (*Comment l'esprit vient aux bêtes*, 1997, *Les animaux pensent-ils?* 2003), agency (*La nature de la volonté*, 2005), and epistemic self-evaluation (*The Philosophy of Metacognition: Mental Agency and Self-Awareness*, 2013). From 2006 to 2009, she led an interdisciplinary ESF research program on metacognition as a precursor of self-consciousness (*Foundations of Metacognition*, 2012). She is currently conducting an interdisciplinary project funded by the European Research Council, entitled "Divided Metacognition: when epistemic norms conflict". Its aim is to provide a naturalistic account of epistemic norms and of the associated norm awareness in human children and adults from different cultures.

Timothy Williamson has been the Wykeham Professor of Logic at Oxford University since 2000. His publications include *Identity and Discrimination*, *Vagueness*, *Knowledge and its Limits*, *The Philosophy of Philosophy*, *Modal Logic as Metaphysics*, *Tetralogue*, and about 200 academic articles on logic, metaphysics, epistemology, and philosophy of language. He is a Fellow of the British Academy and of the Royal Society of Edinburgh, Foreign Honorary Member of the American Academy of Arts and Sciences, Member of the Academia Europaea, Foreign Member of the Norwegian Academy of Science and Letters, and Honorary Member of the Royal Irish Academy. He has held visiting positions at MIT, ANU, Canterbury University (NZ), Princeton, UNAM (Mexico), Chinese University of Hong Kong, University of Michigan, and Yale.

Index